Women in Engineering and Science

The Springer Women in Engineering and Science series highlights women's accomplishments in these critical fields. The foundational volume in the series provides a broad overview of women's multi-faceted contributions to engineering over the last century. Each subsequent volume is dedicated to illuminating women's research and achievements in key, targeted areas of contemporary engineering and science endeavors.The goal for the series is to raise awareness of the pivotal work women are undertaking in areas of keen importance to our global community.

Anna Kelly

Good Work

An Oral History

Second Edition

Springer

Anna Kelly
Portland, OR, USA

ISSN 2509-6427 ISSN 2509-6435 (electronic)
Women in Engineering and Science
ISBN 978-3-032-27546-2 ISBN 978-3-032-27547-9 (eBook)
https://doi.org/10.1007/978-3-032-27547-9

This Springer imprint is published by the registered company Springer Nature Switzerland AG
The registered company address is: Gewerbestrasse 11, 6330 Cham, Switzerland

Acknowledgments

Hilary Boudet has supported every research project I've ever done. I left her tutelage years ago, but she still answers every email, still gets on the phone when I need her. Without her, I wouldn't have the confidence to try half of what I try.

This project wouldn't exist without Sally Duncan. She ran the Oregon Policy Analysis Laboratory, saw something I'd written, and called me. During that conversation, she confessed she didn't see any impactful work being done on climate change. Hearing that from someone I respected, changed everything. That's when I knew I had to do something.

Thomson Comer built the tools that made this work possible. Elizabeth Savage gave me space and took me on vacation so I could actually focus. Both of them believed this was worth finishing.

Patrick Kelly, Sandra Comer, Craig Sinnamon, Tiffany and Evan Hatteberg, Elayne and Ivan Kuletz, Michael Callahan, Colm Otten, and Kathleen Belkhayat provided edits and made this book better.

To everyone who agreed to be interviewed: this isn't my work, it's yours. Your willingness to share your stories made this book possible. Special thanks to Miles Gordon, Edward Maibach, Zachary Owens, Amy Fried, Evan Hatteberg, Greg Wikler, and Alex Osteen, whose interviews informed this book even when space didn't allow their full chapters to appear.

My parents, Sandra and Hugh Comer, let me practice my arguments on them over and over. Their patience shaped this.

Jane Peters and Jill Tietjen helped me find a publisher.

There are many people working on climate change who couldn't be included here. I hope this book honors their work too.

Patrick, Ruth, and Maggie: thank you for your patience, your sacrifices, your love. I couldn't have done this without you.

Competing Interests The author has no competing interests to declare that are relevant to the content of this manuscript.

Contents

Why This Book Exists 1
How to Read This Book 1
The Limits of This Book 2
What I Was Trying To Do 2
Why Anonymize 3

Part I Work With Your Hands

The Solar Plant Manager 7

The Crane Operator 11

The Solar Installer 15

The Solar Developer 19

The Solar Trainer 21

The Wind Development Executive 25

The Wind Farm Manager 29

Part II Work With Data and Science

The Engineering Professor 35

The Geological Engineer 39

The Energy Data Scientist 43

The Climate Science Student 45

The Behavior Researcher 49

The Energy Software Developer 53

Part III Work With People and Policy

The Campus Organizer 59

The EV Policy Specialist 63

The Case Study Curator 65

The Federal Energy Official 67

The Forest Climate Manager 71

The Utility Partnerships Manager 75

The Energy Access Advocate 79

The Watershed Manager 83

Part IV Starting and Building a Climate Business

The Program Designer 87

The Methane Capture Entrepreneur 91

The Climate Investment Educator 95

The Energy Consultant 99

The Capacity Builder 103

The Workforce Developer 107

The Program Evaluator 111

Part V Building the Grid of Tomorrow

The EV Program Manager 117

The Charging Network Manager 119

The Microgrid Consultant 123

Part VI The Efficiency Ecosystem

The Industrial Efficiency Engineer 127

The Public Buildings Manager 131

The Energy Program Leader 135

Part VII The Long Haul

The Wind Turbine Technician 141

The Utility Policy Manager 145

The ESG Consultant 149

Appendices 153

Index 163

Why This Book Exists

The International Energy Agency projects 14 million clean energy jobs globally by 2030. That number seemed wild to me at first, but the more people I talked to, the more I thought: yeah, maybe. I didn't write this book because of workforce projections, though. I wrote it because I have a daughter, and I want her to grow up in a world where humans aren't spending all their ingenuity just trying to survive what we've already set in motion. Climate change threatens food security, political stability, basically everything we've built. Every year we delay is another year of wasted human potential. Brilliant minds solving problems we created instead of problems we choose. I want my daughter to grow up dreaming about being a starship captain, not worrying about becoming a climate refugee. These 37 people are already doing the work. This book is my attempt to help you join them.

These voices come from across the country and across the political spectrum, from urban California to rural Texas, from progressive nonprofits to conservative farming communities. Climate work doesn't belong to one party or one coast.

How to Read This Book

You don't have to read this straight through. If you're a student wondering what climate careers exist, start with Part "Work with Your Hands". The installers and technicians will show you what entry-level work looks like. If you're a data person curious about the analytical side, Part "Work with Data and Science" has your people. Mid-career professionals looking to switch fields should head to Part "Starting and Building a Climate Business", where founders and consultants describe how they made the leap. If you care about policy and systems change, Part "Work with People and Policy" has organizers, advocates, and federal officials working through bureaucracies. Part "Building the Grid of Tomorrow" covers grid infrastructure, Part "The Efficiency Ecosystem" is energy efficiency, and Part "Work Through

A. Kelly, *Good Work*, Women in Engineering and Science,
https://doi.org/10.1007/978-3-032-27547-9_1

Teaching and Innovation" features teachers and innovators who are building the next generation's capacity. Read the parts that call to you. Come back to the rest later.

The Limits of This Book

Thirty-seven interviews is not a labor market study. I found interviewees through snowball sampling: one person led to another, who led to another. That means these stories cluster around my starting points: renewables and efficiency, the sectors I knew best. They skew coastal and western, where much of the early climate workforce emerged. The people who agreed to talk to me are, by definition, people who stayed in the field long enough to have something to say. That's survivorship bias. You won't hear from people who tried climate work and left, or who wanted to enter but couldn't find a foothold.

I also didn't interview anyone from the nuclear industry. In 2020, when I began these conversations, I was looking for technologies that were actually being deployed, paths that could happen given American politics. Nuclear wasn't one of them. Not because it doesn't work. France proved that decades ago. But climate change, by itself, wasn't urgent enough to overcome American resistance. So I focused on what was actually happening: solar, wind, efficiency, electrification.

That's changed. The demand for AI is doing what climate couldn't, making nuclear politically viable. For the first time, the political economy lines up. But I can't go back and add interviews that don't exist. These 37 stories represent the climate workforce as I found it, in the places I looked.

This book captures what the work *feels like* for those doing it, not a complete map of every climate career pathway.

What I Was Trying To Do

I wanted to capture what it feels like to work on climate.

Not the policy arguments. Not the science. The *work*. The daily experience of being someone who gets up in the morning and does a job that happens to matter for the planet.

When I was trying to figure out what to do with my career, that's what I couldn't find. I could find projections of how many solar jobs we'd need by 2050. I couldn't find someone telling me what it felt like to install solar panels. Whether you got bored. Whether you felt like it mattered. Whether your family understood what you did.

These interviews are my attempt to answer those questions.

Why Anonymize

The early drafts used real names. Some people weren't happy with how they sounded on the page. Transcripts can be unflattering. People say things in conversation they wouldn't say in a formal interview. Editing choices that felt right to me felt wrong to them.

For the final version, I made everyone anonymous. No names, no identifying companies, no specific locations. Just "A Wind Farm Crane Operator in the Great Plains" or "A Microgrid Consultant in the Pacific Northwest."

The stories are true. The names are gone.

Part I
Work With Your Hands

When I interviewed the solar installers, the crane operators, and the turbine climbers, I kept noticing what they talked about first. The wind. The views. Heat.

Two of the installers are military veterans who found this work after the service. A Navy sonar technician who says "sure, solar, sonar, it's just one letter" and figured the rest out. A Marine who replaced diesel with photovoltaics and never looked back. The other, a marine who came home from Afghanistan and replaced diesel generators with solar panels. He calls the work "Bolts and Volts" and wishes high schools taught a class by that name.

A man remembers his first climb up a 300 foot tall wind turbine, his knees shaking, and his arms clutching the ladder. When he moved into management, he was asked how many people could die on his watch before he would resign. Everyone had to answer, and everyone remembers what the others said. Another wind expert balances environmental needs, demonstrating how conservation really works.

The residential solar installer is the only woman in her entire field. She scrapes her knuckles on asphalt shingles all day. She installed solar on her parents' house over Thanksgiving while her dad kept climbing the ladder to help. "Thanks Dad."

Each of the seven people who told their stories in this section has a story that reminds me of Working by Studs Terkel. These jobs have wind, danger, and tired feet at the end of the day. As someone who has spent their entire career working with computers, I feel like there's probably nothing in the world.

Some of these people stayed in the field. Some ended up moving to office work, to software, and to training. Careers don't move in straight lines. But they all started by climbing something.

What do I want you to feel after reading these seven chapters? I kind of hope you'll decide it's worth spending a summer on a roof so hot, it melts through your shoes. I kind of hope you'll want to climb a turbine.

The climate crisis needs people willing to work on it. That's it. So I hope you want to work on it.

The Solar Plant Manager

West Coast

I've got 250 megawatts of renewable power coming on every day that is supporting hospitals, grocery stores, and other important places. I want the world to quit trivializing it. No matter how many gigawatt hours are coming on, the media just doesn't think it's good enough.

I also don't understand the divide between working-class, blue-collar folks in conventional oil and gas, thinking renewables are going to take their jobs. I couldn't see that as being further from the truth.

At my plant, I have 1,032 solar trackers, but they use grease made by Mobil. I need petroleum products for my solar plant to run. Even if I didn't, we can't just switch from one fuel type to another. It has to be a slow transition. My plant is located in a central California county, about halfway between LA and San Francisco. It's a 250-megawatt plant developed with a grant from the DOE.

As far as I'm concerned, it's the happiest place to work. I love going to work because my plant is outdoors. I get to work outside all the time. I never have to worry about my hamstrings or lower back hurting from sitting at a computer too long.

The modules are very spread out. Some of the land is migration land, so we had to buy additional land to be managed by a conservation agency.

After a good rain, it's got the most beautiful wildflowers you can imagine. Look up a California super bloom. Unless the flowers are in bloom, I won't see a single car on the roads I'm driving on for work. I work on everything at the plant. At the module level: 425-watt modules, wired in strings of 10, that we combine into 18 rows that feed into inverters. I get to work everywhere from the module up to our point of interconnection with the utility.

A. Kelly, *Good Work*, Women in Engineering and Science,
https://doi.org/10.1007/978-3-032-27547-9_2

Our whole job is to keep assets at a high rate of availability. The energy is promised via a power purchase agreement. My whole goal is to keep equipment up and running.

Most of my work is based around a preventive maintenance program, mixed with corrective maintenance, meaning unplanned things that take an asset offline. I've got one of the best jobs in the world. I work four 10s. This week I was doing validation of our drone flights. We use unmanned aerial vehicles for aerial thermography, looking for module anomalies, hotspots, diode failures.

We do this for safety first. Hotspots can cause thermal events. Fires. We don't want to cause fires. Second, we want the healthiest PV generator we can have.

Every couple of years, we get to do the fun stuff: really big maintenance on the substations. The substation is the critical component of renewable energy. We give our technicians the ability to master as much as they can. That way we can take pride in what we do.

We are a continuously learning crew. I feel very fortunate. I get to go hang out with my friends every day and learn. Even though it's good, we don't have enough workers. The labor force just isn't there. There's a lot of job hopping. If you want to be, you can start out as a high-performer and theoretically be a site manager in 2 years. There's so much new capacity coming online that there's lots of opportunity.

In 2012 and 2013, I was working as a generator technician in the military, working on diesel-powered equipment in Afghanistan. When I came back, the only job I could find paid $12 an hour. PV has afforded me a great quality of life. I like replacing all the diesel equipment with PV. I worry the labor force isn't developing the skillsets needed. From my perspective, what you need is a good attitude, common sense logic, the ability to troubleshoot electricity, and some logical thinking. I don't see that skillset in new people when I'm looking over resumes.

I think high school students are good candidates. People need algebra, but I don't need anybody doing calculus. What I really need is somebody who played sports. Teamwork is a big thing at these plants. You've got to work with different personalities toward a common goal.

Other than that, it's about personality and showing up to work. Getting people to show up is a big problem. These plants are in very rural areas with no local workforce to draw from. I try to advocate for people with diverse skillsets. I remind my hiring managers: you have to take a chance on somebody. As long as they've got a good attitude, they'll be able to do it. We're not SpaceX. We're not trying to go to Mars. We need to use a multimeter safely. We need to work around high voltage that does have the potential to kill us.

When I got out of the Marine Corps, I couldn't conceptualize what I could do. I thought I had to be an aviation ground support equipment electrician. I didn't realize that if it had a bolt, then I could play with it. That's my thing now. Bolts and Volts.

If they had a high school class on Bolts and Volts with a bit of computer networking, it would be killer. I think it's the responsibility of the manager to remove barriers that rob employees of pride in their work. That's stolen from Dr. W. Edwards Deming, the quality guru who went to Japan as part of the Marshall Plan after WWII. I think I'm making an impact, but it's a strange concept. I like that I produce energy. I like that it's used in my country. I try not to think I do too much all by myself. The biggest impact I have is on a team level, not necessarily the giant world.

We give donations during Easter and Thanksgiving. I feel good that the turkey my neighbors are eating is from my company. I view myself as a maintenance technician, not a renewables person or a solar person.

When I look at the renewables industry, I feel like we're making a lot of Yugos and not a lot of Toyotas. There's a lot of incentive for short-term financing for project development. But I don't care how much capacity there is. I don't care how many Yugos got made. What I want to know is how many are still on the road today.

If you build a 600-megawatt plant but use tier-two modules and inverters, you're going to have 75% energy availability after 5 years. That doesn't help us.

I've heard of plants staffing 200 megawatts per technician. That means one person covering a 200-square-mile area. I get to be proud of having 99.9% availability. People who are stretched feel happy with 80% because they're running around like chickens with their heads cut off.

The way the industry is being run isn't protecting America. We need to improve our quality so we can be competitive.

I'm worried that America is simply behind on solar.

The Crane Operator

Great Plains

I am fifth generation in the crane industry. Back then, until 1989, cranes weren't involved in renewable energy. It was all petrochemical. I started out as a rigger, using cranes along the way, Egyptian-style rigging, moving heavy equipment and transformers.

Every time I start to tell my story, I go back to being at my great-grandfather's house. There were pictures of huge mule trains pulling transformers off barges. They'd create a derrick with guidelines. That's how it started and progressed to where we are now: 2,200-ton hydraulic cranes that can drive down the road. I was an operations manager at one company for years. When I turned 50 and the owner died, his brother died, and his son died, I decided to take on a new challenge.

Then I got a call from my current company. I didn't want to go to work anywhere. I was going through that with my mom. They insisted on coming to see me across the street from the hospital. Then they paid me for 2.5 months before I ever went to work with them. My wife called and asked if I had signed up. They said they knew I was going through a tough time and felt an obligation to stay.

They are the largest crane company in America. Now my job is technically sales. I worked as an operations manager for years, but I didn't sleep a lot and the stress was killer. I had a good time doing it, but now I go out and sell the things without the responsibilities.

I used to have to be directly involved in dealing with personnel every day, dispatching crane trucks, managing folks. I was running safety, trying to find ways to keep sites safe, but I missed my kids growing up.

A. Kelly, *Good Work*, Women in Engineering and Science,
https://doi.org/10.1007/978-3-032-27547-9_3

When my son was 12, he came up to me when we were getting ready to go on vacation. He said he wanted to talk to me. Then he grabbed my briefcase and put it on the table and started crying and said, "Can't you just leave this home?!"

It was that kind of stress. I never stopped working. You look back on your life, and you don't want to call it a mistake. But as you get older you start to realize what's really important. I've dialed back a lot. We're working on wind farms. They might be hundred-meter towers. A majority of what we're doing is maintenance of the turbines. After they're up and running, they break down sometimes. Gearboxes and lots of moving parts. When those parts go bad, the farm has to shut the turbine down and lose revenue. Then we come to the rescue.

We run our cranes and get the rotors dropped, take the rotor off a turbine and bring it to the ground. Then we'll have a secondary crane that hooks onto the blade pointing downward, and we trip it so it comes down horizontally. We set it in a stand, pull the lid off, pull out whatever components are broken. Exchange the components and put it all back together. If I see a request with a term like "turbine deconstruction," that lets me know the turbine broke a blade and is damaged. Problems like a dent in a tower can lead to catastrophic issues. Turbines can go out of control and explode and tear themselves apart and fall to the ground.

There are lots of videos on the internet. It's gnarly. It's scary.

I've been onsite when a turbine has thrown a blade. That's when the whole thing gets out of balance and destroys itself. It's an amazing thing to watch. It's scary being there.

One time there was a wind storm near Palm Springs. One of the blades whipped off and landed in the ground, not on the ground but IN the ground like a spear, a hundred yards ahead of us as we were driving. Oh my God. Then the whole turbine was out of balance and just tears itself apart.

In those cases, we go out to rebuild and tear down the whole tower. I enjoy going out to the sites. Most sites are in cool places. Desolate, but beautiful. It's awesome to be outdoors and not stuck.

I drive out into the desert and start climbing up into a valley. As I climb in elevation, it turns into forest: deer, bear, bobcat, mountain lions. Crazy and weird. In the middle of the desert. Then it's all crystals and rocks, and there's a great energy up there. I go to the top of the hill and kick back and read a book and chill out.

Most folks I work with feel exactly the same way. They put wind turbines in places that won't bother people. They put them on farms because farmers like collecting the cash.

The farmers like getting checks every month. The risk-reward works out for them. If a young person wants to do this, they just need to call up a crane company and let them know they're considering this line of work.

I've had plenty of young people come up and sit with me and tell me their dreams. If you're passionate and have a good work ethic, we'll bring you in as a yard kid. Then you get to see things from the outside as it's happening. You learn rigging and components and see how the business operates.

Typically, folks are yard kids for a couple years; then they'll send you to an apprenticeship program through the local union of operating engineers. At that point, you step into a career with really good money. Crane operators make a couple hundred grand a year if they're nice to work with.

I can tell you this job is never boring. I've worked in movie studios, military stuff I can't talk about. It's not glamorous, but it's cool.

The Solar Installer

Mountain West

To install residential solar projects, you meet at the warehouse every day at 6:30 a.m. and get your assignment. We might have a heads-up that the job is far away and could take up to a week. Then I get into the truck with my crew and head off.

I did lighting in set construction for theater in college, so I was comfortable with heights, tools, and electrical work. The first thing important to know is where the solar is going. Some roofs are multifaced and panels are going on multiple surfaces.

I show up and one of us sets up a ladder and installs the safety equipment. We use hitch clips that slide underneath the top row of shingles and get screwed into a rafter. Our safety ropes clip into those, and the ropes clip into our harnesses.

Once safety equipment is installed, we go up and draw out where the array will go. We measure everything, make sure it's centered, make sure everything fits within the plans.

Then it's time to install: footings for the rails, then the rails, electrical, and then we start slapping glass.

It's important to get them square. That's the real trick. Someone will be on the ground yelling up, "tip it this way, oh now go back a little bit." Once the first few panels are square, the rest works out. The actual work of installing residential solar is painful and messy.

You're working on your hands and knees on asphalt shingles, which are the roughest grit sandpaper. All day long, you scrape your knuckles and knees. Don't forget we do a lot of these installations in summer, and the roof gets extremely hot. Too hot to touch.

Residential solar will always be on the sunniest part of the roof. Sometimes I get to the point where I think, Jesus Christ, can I just get into the shade for a minute?

A. Kelly, *Good Work*, Women in Engineering and Science,
https://doi.org/10.1007/978-3-032-27547-9_4

But no. If I'm doing my job correctly, I will never be in shade. I will be in the sunniest part of the roof all day long.

Then if I leave for lunch and come back, the panels themselves have had time to heat up. Touching them results in second-degree burns just from making contact with the hot metal.

We wear t-shirts but never shorts unless we've got kneepads. Tennis shoes, and we get a budget for them from the company. The roofs just eat your shoes. We all go through several pairs a season.

I do installations in Minnesota and Wisconsin, and the season lasts a long time. By winter I'm not up there in tennis shoes anymore but in snow boots. And we aren't laying rails first. We're shoveling snow off the roof. The wind is rough regardless of season. The panels are a 55–65 pound sheet of glass. I'll hold this $300 thing in the crook of my arm and balance it on my head and climb the ladder. As soon as I clear the house, it turns into a kite.

I personally have never broken a panel. I've seen people drop them, bolt them down too tightly and crack them, and accidentally step on them. One coworker was trying to maneuver to an inaccessible area and stepped just a little too far from the frame, and the whole thing broke into spider webs in an instant. I put solar on my parents' house in Iowa. That was fun. It was through my company. I did everything on the roof and ran all the electrical. I did it over Thanksgiving.

My dad kept coming and standing on the ladder to help and ask if he could hand panels up. Thanks, Dad. We had a nice time. He even came up on the roof once or twice. The crews are all young men. They're full of beans. They're really hungry to have these jobs.

I'm in my early 30s, a decade older than most people I've worked with.

I try not to let being a woman stand in the way of doing whatever job I want. There are some limitations I face. But the guys I've worked with have been understanding when I can't physically do a thing. Sometimes I have to ask for help. Sometimes I don't have the upper body strength for certain aspects.

It's really exhausting to pull my weight on a crew with very, very fit 20-year-old men.

We have to get more women in the game. I'm the only woman in my field. This job really takes a toll physically.

All day you're standing on an angle. The harness pulls at your body, your hips. You brace yourself all day. I always wear holes in the same spots in my shoes. My left shoe is where the ball of my foot hits the roof. I think I use it as a push-off point all the time. People stare and watch us work. Especially in suburban neighborhoods when you're there all day. The bus lets out, people come home, kids start playing, and then people gather and watch. You end up being a spectacle.

People are very interested in how it happens. I've done houses right next door to ones I did before later in the season. Solar is a surprisingly anti-union industry; I think because it's a transient workforce. For many people, they just work part of a season. Except for electrical positions, you can easily learn the job in a short amount of time. Those jobs don't require the training that other trades require.

I don't know too many people who can do it for more than 5 or 6 years. At a certain point, you just start to break yourself.

I like it though. It gets me outside and keeps me in shape. Until I find something better to do, it's a job and not a career.

I think it's important we have this conversation. The anti-union stuff is atrocious. There's no ethical consumption under capitalism… so fuck it all? But we have to start somewhere. It's not perfect. There's no perfect solution.

It's not like we can't try to make things better. We have to accept that our solutions won't be perfect immediately. This technology is incredibly promising, and we have it now.

I'm sure we have a long way to go before we make even our most renewable energy as clean and safe and fair as possible. But we have to start somewhere. We have to at least try.

Otherwise, we'll just lie down in front of the pipeline and whatever. Just take me. It all sucks. Yes, it sucks. But this sucks less than the alternative.

The Solar Developer

Upper Midwest

We're a construction developer that also does solar gardens. I knew the CEO from way back. I wrote him on LinkedIn when I was perfectly happy at my other job, just to connect to play beach volleyball. He asked if I wanted to work at his company. He made me an offer I couldn't refuse.

My company kicks ass, honestly. I'm going to law school at night, and they support me. The main focus of my job is portfolio administration. There're a lot of different things to get a solar project off the ground. I started as a policy analyst. I go to solar energy industry events and lobby days in DC where you have discussions with representatives or senators.

I always enjoy those. They're a good way to meet people. I work in financing for distributed generation: commercial-scale solar arrays and solar gardens. We're intermediaries between the people that are going to own them.

Each project is its own little company. You have to get that company registered as a power plant: little sun-powered power plants.

I've taken on more project monetization recently. We don't want to hold a project for very long after we buy it. It's a lot of risk to hold.

Instead of dealing with venture capital or high-net-worth individuals, we're dealing with banks, a whole different process. It's all about showing due diligence. I arrange that the due diligence gets done correctly; then I'm trying to take it to a bank to make money.

The life of a middleman: trying to hold this hot potato for as little time as possible. It's a strange situation with bottlenecks everywhere. We have been trying to close a three-project portfolio since November; then December came, then January,

A. Kelly, *Good Work*, Women in Engineering and Science,
https://doi.org/10.1007/978-3-032-27547-9_5

and now end of February. We're trying to get this dang bank to take these projects. I kind of fall apart a little bit.

The whole due diligence process is about appeasing the bank. There's all sorts of real estate stuff. Mineral rights. These are huge parcels, like 30 acres. Permitting, environmental documentation, reliance letters, checks on the checks. If it takes too long, we have to renew environmental reports because they get too old. Six months is pretty short. We try to do greenfield development: the land is just a field and no one else has thought about doing a project there.

In rural areas, greenfield developers go to a farmer and offer a monthly fee for their land. Then there's always a hubbub. The neighbor is like, "I don't want a solar garden in your backyard." And the farmer is like, "Fuck you, I do what I want. I want a solar garden that's going to pay me money every month." There's this whole crazy system of rules for solar gardens.

Community solar gardens require a minimum of five subscribers per garden. Usually one anchor tenant at 40% fills in the rest with others. You need a mom-and-pop shoe store, a coffee shop, a brew pub. You can't have two solar gardens within a mile of each other in some places.

There are asinine rules that utilities put into play. The worst is the county rule. You have to get subscribers from contiguous counties. The county lines in some states are crazy. One county might be just slightly off from another. You can't use that subscriber, even though it's a huge bottleneck.

It's all part of the same grid. I don't see why not. But no. That's how it goes.

The Solar Trainer

West Coast

I rode nuclear submarines as a sonar technician for 20 years and retired at Pearl Harbor. I got out in 2007, which was not the greatest time to transition. Everything was tanking. I got to San Francisco with my wife and four kids and started calling companies. They all said they didn't have anything.

My specialty was being a trainer, and no one wanted a trainer. I ended up working part-time jobs: one at REI, one at IKEA.

It wasn't long until a solar company called, saying they needed a trainer. I didn't know anything about solar. Because of my work with sonar, I did know electronics. And I thought: sure, solar, sonar, it's just one letter. I'll give it a shot.

The company had been in business for 30 years, going all over the world installing off-grid systems. They'd gone to the middle of the Pacific to an atoll and installed a 400-kilowatt system. For the people who lived there, they now had electricity that didn't run on a generator.

It was so cool. I'm getting goosebumps right now just talking about it.

I've been in the industry ever since. After 7 years, I went to another company that just did storage. They'd been working in Germany with lithium-ion batteries. My previous company was all lead acid.

We beat Tesla to market. Of course, he gets all the publicity. But we were a little startup, and I'm proud of what we did. I trained a lot of people.

Then I moved to another company that did module-level solar installations. There had been pushy salespeople shoving solar down people's throats even if their roof couldn't accommodate it. The rule of thumb was to never install solar if there was roof shade between 9 a.m. and 3 p.m. A lot of companies disregarded that.

A. Kelly, *Good Work*, Women in Engineering and Science,
https://doi.org/10.1007/978-3-032-27547-9_6

Now we're feeling the effects: homeowners struggling with abandoned systems from companies looking to make a quick buck and get out. That's starting to happen with storage, too.

I opened my book with a true story from the California wildfires: a homeowner who spent a lot of money on a solar and battery system for resiliency, and they had to leave after only 2 days.

In my experience, the breakdown happens in presales because of a lack of education. That's why I wrote my book.

Salespeople come with a basic pre-made sales sheet with green check marks and red Xs designed to make their manufacturer look best. The spec sheets don't actually allow for comparison. How can customers make educated decisions? There's so much disparity in what's available. No perfect solution, my company included.

One of the biggest things I see is a lack of surge power. It's not smart to run your three-ton air conditioner when the grid is down. I get that people live in Phoenix and Tucson and it's 120°. But not everyone has room for a big enough battery to run a big air conditioner during an outage. That would be a whole garage worth of wall space. People think if they have a system with a certain number (8 kilowatts, 11.4 kilowatts), they'll get that much power all the time. But it's not true. The output is limited to what the battery can provide. Solar is just gravy, and people don't understand that.

Another thing that confuses people: the difference between kilowatt and kilowatt-hour. A kilowatt is how much power. Kilowatt-hours are capacity, the number on top of your phone telling you how much battery you have left.

But nobody cares what a watt is. They know how much they pay. You have to use anecdotes and analogies. It depends. It always depends. People say they just want to run their laptop, charge their phone, keep their refrigerator going. Americans go big, but we should help people make good decisions, like not putting their air conditioner on an essential load panel. If you're worried about resiliency, put a mini-split in the living room and call it quits. Otherwise your system gets real big real quick. Discharge matters too. If you have a 10-kilowatt-hour system but it's only rated to 90% discharge depth, you don't actually have 10 kilowatt-hours. You've got nine. Plus the designers bake in headroom. You're probably getting 5–7 kilowatt-hours. At the grid level, there are interesting things you can do with batteries. Sometimes there's too much solar, and batteries can suck up the excess. With electric vehicles, companies are promoting EV-to-Home systems.

I call them rolling dump loads. The idea is you can have your electric vehicle power your house.

The batteries in EVs generally have 1,500 to 2,500 cycles. Every power outage where you use EV-to-Home, you're decreasing the battery lifespan. Contrast that with lithium iron phosphate batteries used in stationary storage. Those have 6,000 to 10,000 cycles. Take 10,000 cycles; you've got 27 years.

Don't mistake cycle count for warranty reliability. Think about your cell phone, your laptop. A brand new one has an 8-hour charge, but after a couple years you have to keep it plugged in. I've had a tremendous impact on this industry because I've always been a technical trainer. I've talked to more people than a salesperson could ever hope to.

In the heyday of my career, I was training 4,000 people a month.

In an industry that is small, I've made a name for myself explaining how this stuff works, just being honest and upfront. I cut out the bullshit and talk, and people appreciate that.

Ever since I wrote my book, I go to solar shows and people tell me they love it. When I develop a certification program for a company, service calls go down. It's a proven system.

The impact is hard to quantify, but I get a lot of pats on the back. People buy me beers every once in a while. I'll take it.

The Wind Development Executive

Texas

We're an independent power producer, what we call an IPP. We only do zero carbon emission project development. We work across North America as cradle-to-grave developers. We originate projects from scratch: finding landowners, making the interconnection request to the utility grid, getting permits, sourcing equipment.

We work across wind, battery storage, and other emerging renewables. We do engineering, procurement, construction, commissioning the project, then operations and maintenance. When I say cradle-to-grave, we don't just do one phase and turn it over. We run the whole gamut.

I work in grid-scale power. These are utility-size power plants. We're basically replacing the old infrastructure that ran off nuclear and coal and gas-fired plants with renewable energy. I've been here 13 years. I'm a senior development director. I oversee a portfolio of early and mid-stage projects across the Western US: Colorado, Arizona, New Mexico, Utah, some Pacific Northwest.

I'm also overseeing acquisitions of projects from smaller developers, projects with a land lease or interconnection application but without capital to actually build. These are very capital-intensive, \$300–\$500 million.

My day-to-day is managing all the early-stage aspects. We prospect areas close to grid interconnection points with low conflict with natural resources: environmental species, water features, archaeological sites, scenic view sheds.

Wind projects, on average, have a lifecycle between 5 and 10 years. Long-term endeavors. Not a quick turnaround. We find out what to build by looking at utility plans. All utilities put out integrated resource plans showing a 10-year outlook. They explain their load forecast, current energy mix, and procurement targets for

A. Kelly, *Good Work*, Women in Engineering and Science,
https://doi.org/10.1007/978-3-032-27547-9_7

renewable energy. We look at those plans and decide where projects are likely to be built, how big, and when.

Further out, utilities tell us where they're building new mega transmission lines. These take 20 years to design, permit, and build. They encourage developers to prospect projects to plug into these new lines. We're looking at those roadmap signals. There are commercial and industrial players driving a lot of renewable development right now, desiring to source all their energy from renewables with a guaranteed energy price. The beautiful thing about a renewable project is it comes with a fixed price over time. You're hedging against natural gas volatility. If you lock in a renewable power purchase agreement at a certain rate for 20 years, that's the price you pay. For wind and solar, you need a good resource. We're looking for windy places but not high-use areas for Golden or Bald Eagles. We don't want to impact views from the Grand Canyon or culturally sensitive areas. There are radar issues, airport issues, and military training routes.

I got the first-ever Golden Eagle take permit from US Fish and Wildlife in 2015 for a project in one California county. It authorized the project to take more Golden Eagles than we ever will, which is good. The service wants to be conservative.

Because it was a no-net-loss policy, we did power pole retrofits in areas where Golden Eagles were being electrocuted on dangerous transmission lines. We paid for retrofitting. As a result, more Golden Eagles have been saved than have been endangered by our spinning turbines.

I did other endangered species permitting where we set aside, in perpetuity, endowed lands for desert tortoise. They could have a nice preserve for the rest of their lives. In exchange, we got approval to build solar on land with marginal habitat. There are all these constraints and you fit a puzzle piece on the map with the least impact and greatest purpose. We send land agents to knock on doors. "Hey, are you interested in being part of a project?"

We work with local jurisdictions and do economic benefits analysis. We have community engagement meetings where we speak about who we are, what we're trying to do. In the case of wind, the final footprint takes about 3% of the overall land we lease. For the rest, they're free to do what they used to do: farming, grazing. Those activities coexist without any problem. It's a good investment for landowners. Sometimes it allows them to keep land together they otherwise might have to sell.

Solar is more land-intensive because it covers the land with infrastructure. It's a conversion of land use type, but we have decommissioning plans to restore the land to its original state once the project has run its life. Even so, there are many factors that can stop a project. Better to find out earlier than later. I keep my eyes wide open and don't chase projects with low likelihood of success.

We've seen interconnection expenses too high, infrastructure upgrades too expensive. Communities are coalescing around not wanting projects. Rural farming communities are not wanting conversion of irrigated farmland. There are conflicts with military training operations.

There are folks who just want panels on rooftops, not in deserts or adjacent to parks. The reality is all projects, even renewable projects, have downsides. You can't produce power without some residual impacts. You try to do it in the most responsible manner possible. What we'd like to do is bring economic benefits to areas that have been ignored or taken advantage of. If we build a project in their community, there's immediate benefit from construction jobs, then long-term benefit from tax revenues and landowner payments.

In some jurisdictions, we set up tax abatement plans where funds go directly to school districts. We did a project in rural New Mexico where we almost doubled tax revenues for certain school districts. Pretty amazing.

We've done educational programs, sponsored community college programs, paid internships. We had a woman in southern California who had an internship doing O&M on a wind turbine. We hired her and that's her career now.

We look at it as a long-term relationship. When we build a project, it could have a life of 40 years. It's important the project is accepted locally and provides benefits. On a grander scale, it's a transformational change in how we produce energy. Every step we take to get another polluting project offline and to green our grid gets us closer to energy independence, which is pretty important right now with all the geopolitical issues. Our outlook is extremely positive. Can we do this fast enough to combat climate change disasters? I'm not sure. We're going to need technological breakthroughs. Maybe carbon sequestration.

But I'm very optimistic that there's no way to stop continued growth of renewable energy because now it just makes economic sense. With volatility in energy markets and the true cost of carbon someday priced in, renewables will look like an even better deal.

We got renewables to price parity. There will be no way to stop what's going to happen. We will make this transition.

Will it be fast enough? I don't know. We're all shocked by how quickly this nightmare of climate change has come to our doorstep.

It's here. It's here to stay. It's scary. I don't know if there are going to be climate refugees or what. But it's real, and unfortunately it is happening faster than we predicted.

The Wind Farm Manager

Great Plains

I work in the wind sector, providing clean energy from wind turbines. When I was little, I remember always staring at the wind parks. I felt you needed extensive schooling and training. I didn't even go to college. I graduated high school and went straight to work.

I was helping out my father-in-law laying sod at a client's house. As we were laying the sod, this gentleman came out and observed. He said in a stern voice, "Hey, can I talk to you?" I walked over thinking he was going to complain about our job.

Then he sat me down and I had a soft interview.

I didn't even know a turbine was called a turbine. I called it a windmill.

But I joined at a very entry level and worked on turbines. I was fortunate to jump into management. I'm directly responsible for managing the site and the assets. One of our sites has 70 wind turbines for a total of 140 megawatts. The turbines are 2 MW each, 80 m tall, a little under 300 ft. The footprint covers about 5 acres, all owned by individual parties.

Having turbines on land owned by others is tricky. Different landowners require different things. If we don't keep them happy, they can default on a contract and say they don't want our turbines anymore.

To keep them happy, we have to keep the area clean. We're a green company. We try to disturb the land as little as possible. There are designated roads to get to the turbines. If you take a shortcut, you could damage crops: corn and wheat fields. We work in mostly desert areas: cacti, Joshua trees, protected species.

A. Kelly, *Good Work*, Women in Engineering and Science,
https://doi.org/10.1007/978-3-032-27547-9_8

We mitigate for Golden Eagle, scout for Burrowing Owls, and Desert Tortoises. We definitely want to keep our eyes on our surroundings. Every morning we hold what we call a pod meeting. We do planning for the day; it could be a couple minutes or an hour. We do daily stretch and bend exercises before heading out to the fields. It helps make sure everyone is warmed up and is designed to mitigate sprains and strains, which are the most common injuries in the industry.

Safety is a big issue because of wildlife. Sometimes we run across coyotes roaming around the turbines. And we have a vicious and aggressive snake that will chase you and has two different types of toxins. They're very active. Even in the last month or two we've had five or six sightings.

If someone got bitten, we'd rush them to the nearest hospital. There's not an antivenom for that type of toxin. But the chance of survival is high if they make it to the nearest emergency room.

That's the wildlife that's scary. Not to mention the heights. The first time I climbed a turbine was very nerve-wracking. I wouldn't even consider myself afraid of heights. But I'd never been put in a situation where I had to work at heights.

I still remember what my first climb was like. My knees were shaking; my arms were crunched up to the ladder. I was stuck wondering if the equipment would hold me if something went wrong.

I go to the top and climbed into the cell, which is the big box-looking item on top of the tower. Once I was in the cell, I went into the rotor where the blades are installed. The most nerve-wracking part is when I opened the hatch and climbed up on top of the nacelle and looked down and checked my comfort with heights.

It's not slippery at least. It's very textured with strips of grip tape. The maintenance crew goes up following schedules. We climb the towers, inspect mechanical and electrical components to mitigate failures, and change grease collector bags. We make sure there aren't any leaks or protrusions.

The gearboxes are like the transmission in your vehicle. The generator is like the motor. Sometimes generators fail; gearboxes fail. The blades get cracks. They're fiberglass, and we have to go up and perform repairs.

If the blade repairs are minimal, we can have a rope access team rappel to the damaged area. If extensive, we have to bring a crane team to lower the blade to the ground. In my short time here, we've only done two on-the-ground repairs on the 10-year-old park.

Cracks on blades can cause a whistling sound irritating to landowners. We inspect blades every service cycle. We get binoculars on the blades and look for cracks. If we find some, we have drones that fly and inspect for damage. This industry is important. Renewable energy is very important. First and foremost, it provides very well for families, especially in remote areas. It provides really good stability.

I'm thankful it's provided a good livelihood for me and my family. That's the biggest appreciation I have for the industry. It drives me to come to work every day. I have seven direct reports. We do anything from preventive maintenance and troubleshooting to major corrective repairs.

Managing is a skillset of its own. The biggest and most interesting piece is the different personalities. No two people are alike. That's a constant struggle. What makes one employee happy does not make the other happy. It's an interesting puzzle.

I have a mix of people age-wise. I've always struggled with managing people because I'm a lot younger than a lot of folks and came into management at a pretty young age.

One question I've always been asked is how I manage people older than me. I demonstrate that with work. If it's politically correct to say, older generations appreciate it when I'm willing to get my hands dirty. I've always demonstrated my abilities by working side by side with my team. I try to manage and lead from the bottom. That's my model. We turn our ship around by turning followers into leaders. That's my philosophy.

I put together a plan and present it to my peers and direct reports. I explain the why. I want them to drive the plan. If my plan is not the way to go, we sit together. We construct it together. That gives them buy-in and ownership.

I try to coach them, train them, do the most I can to get them to succeed. If I have to let someone go for performance issues, I want them in their exit interviews to know I did everything I possibly could. A few years ago at management training, one question asked of the managers from different states was: what would it take at work to make you resign? What would count as a big enough mistake?

They worded it as: How many people are you willing to kill before you resign?

You'd be surprised at the answers. Some folks said 5, or 10. The numbers were astonishing.

If I were to lose somebody on the job and it had been an avoidable situation, that would do it for me. That would seal the deal on my management career. One of the hardest parts of my work is when someone gets hurt on site. Whether through their negligence or a bad call. I tell everyone they need to be safe because I don't want to call someone's spouse and let them know their husband or wife isn't coming home due to my actions.

The moment your guys walk out the door, it's your responsibility to know right from wrong. I'm here in the office most times. If I'm back in my office and they're out there, I have to rely on their integrity to do the right thing.

Part II
Work With Data and Science

For the most part, the data scientists and researchers in these six chapters stare at the same glowing rectangles I do. They're not climbing turbines or standing on roofs that melt their shoes. But when I talked to them, something surprised me.

They worry a lot.

The PhD student in Montreal doesn't know what comes after his degree. The engineering professor pushes his students to struggle with problems until they break through.

The software engineer building machine learning models says he's in a "supporting role." Other people are doing the real work. He just enables them. The data scientist at the oil and gas consultancy finds hope in engineered plants that sequester carbon through their roots.

One of them dove into permafrost ponds in Alaska wearing a dry suit, crunching herself into a ball to get the air out, collecting sediment samples while someone held her underwater. Her activist sister made her speak at a climate rally. The head of the EPA wrote her and his letter hangs on her refrigerator.

People think data is abstract. Numbers, models, statistics. Who cares about a regression analysis? But a regression tells you how much something matters. It can actually answer that question, which is kind of amazing when you think about it.

These people are the ones who prove climate work is working or not working. There's pressure in that. Being the one who has to say "this didn't work."

But negative results are still results. It's so important to not lie about failure. Especially when you're doing something new.

They're making sense of the world so the rest of us know what to do next. I'm grateful that someone's doing that work.

The Engineering Professor

Midwest

Some people come to engineering not out of childhood dreams of building robots, but because the numbers made sense and the world needed fixing. A lot of our undergraduates want to make a difference. We don't need you all to go be investment bankers. We want you to go do stuff.

I'm a thermofluid scientist. I got into engineering because I was good at math. I was not the strongest student in high school. My focus was on athletics. Engineering was math heavy, so it was a logical choice.

I'm not one of these people who have done Lego robotics since elementary school. It's a harsh reality for some folks who get sold "engineering light" when they're younger and are sold on the idea that it's all fun. Engineering is tough. It's not for everybody.

Some people get that bachelor's because their parents wanted them to, not because it's their passion.

The truth is engineering is a lot of hard work and time alone, crunching numbers, learning how to fail and overcome. I didn't have any illusions. I just thought it would be a good career. Everything I've done has contributed to either energy efficiency or carbon emissions reduction. Public enthusiasm for climate change has increased during my career, but I haven't changed at all. I've been doing the same work all along.

What's come my way has been oil and gas: cleaning up emissions from large engines, turbines, making better gasoline, designing new internal combustion engines. Lately there's been more work on hydrogen. Currently, I'm doing biomass work: wildfires, hydrogen production, cleaning up wood stoves.

A. Kelly, *Good Work*, Women in Engineering and Science,
https://doi.org/10.1007/978-3-032-27547-9_9

Frankly, I would've probably been doing this regardless of global warming consciousness. I'd be doing it anyway.

I've always felt at odds with activists. They make problems worse sometimes by bringing attention without having solutions while I'm quietly in the background working on answers. I do make a difference. In academia, I wear several hats. I'm a tenured professor; I have an administrative role, and I teach.

I teach thermodynamics, which is difficult. I push people hard. My goal is that they learn the subject matter. I'm not there to be friends with my students. I think it does them a disservice. My job is to prepare them for their career. Working with researchers is different than teaching because we're working toward an end goal. I'm the team captain or the coach.

I got my first major federal award as a new professor. I had to build a laboratory and train all my graduate students from ground zero, making sure no one got hurt, all at the same time. That was a great time. They believed in the cause and that fueled me.

Eventually we got undergraduates to train. Some stayed on and primed the pumps, able to train the next group and the next. I try not to be too prescriptive about my hopes for graduate students. I've had some really great ones I'm very proud of. But if they decided to stop engineering tomorrow, I would understand. I would feel like I had fulfilled my commitment that they had all the tools to do whatever they want.

For PhD students, I want them to be able to go into academia, go to a national lab, go to industry, whatever they choose.

I have two students at national labs now. I'm very proud of them. One is doing fundamental studies; the other is administering large energy grants. That is my goal for all my graduate students. My philosophy on teaching is that learning must be painful. Struggling is the pain of learning.

One of the biggest things I teach is how to simplify problems and apply proven approaches.

Students who have never done experiments can be good on paper and have no idea what they're doing in the lab. Things take longer in hardware than they do in software. That's harsh reality.

You have to be methodical if you really want to solve a problem. You really have to boil things down and not listen to all the other noise. One of the most fulfilling things I've seen: An undergraduate student of mine was trying to replicate calculations in a journal article and he was struggling. We were in a group meeting and one of my graduate students said what I was thinking but didn't want to say: "Listen, you have to try to do the calculation. You have to struggle. You have to iterate. This is the beginning. This is how it's done. There's no way around it."

I couldn't have been more proud.

You just have to grind. It's not poetic. It's not sexy. But there are times when it all comes together. Graduate students learn to appreciate their impact more than chasing the biggest salary. I've had them echo back that maybe they'd rather be packing their lunch and wearing cargo shorts instead of having all the flashy stuff.

I didn't get too senior when I was in industry, so I didn't have a lot of people to mentor. As an educator, I've been able to influence so many people. I feel very fortunate. To me, success looks like continuing to support people who are high energy, enthusiastic, and willing to try new things.

It's insulting to the scientific community to think a hedge fund is going to have a massive influence on climate change when there's a lot of smart people working on these problems all the time.

Right now we have a big push for hydrogen, but it doesn't make sense in a lot of ways. We have to create hydrogen. It's not naturally occurring; it's expensive. I encourage healthy skepticism with some of these silver bullets.

I mean, it's tough.

Anyhow, I'm not a philosophy professor. Thank goodness.

The Geological Engineer

Upper Midwest

My grandparents were immigrants, and the work ethic was instilled early. Education was very high on their priority list. I always liked science since grade school.

My undergrad was in microbiology; I have advanced degrees in geological engineering. I work with geologists almost exclusively. From the start I was the token engineer guy in the group.

Engineers tend to think linearly when getting from point A to point B. Geologists have a different approach because we have to make inferences based on limited information. The materials we study are buried underground. We may have a sample, but we also have to rely on cues from the present. I learned early how to make connections from things not necessarily related to what I'm working on.

I started part-time in 1986 and I'm still in the same place. Humankind consumes a lot of resources through mining. People talk about sustainable mining. Some think that's an oxymoron. What drives me: figuring out what to do with the materials removed from the ground.

Not the target materials, not the steel or the nickel. Those represent just a fraction of the entire tonnage. I mean the 95% of other stuff that ends up in stockpiles and tailings. It represents a lot of energy used to get to that point, and it's just sitting there. These metals are necessary for our clean energy future: batteries for vehicles, advanced energy storage. They have to come from somewhere. Not all of it can be achieved from recycling. I read recently that for net zero CO_2 production we would need 40 times what we currently get from recycling.

A. Kelly, *Good Work*, Women in Engineering and Science,
https://doi.org/10.1007/978-3-032-27547-9_10

Since we can't do it with recycling, we need to see what we can do with minerals we're already taking out of the ground. We've patented a pothole repair material using waste materials from mining. It sets up with a chemical reaction like cement that cures quickly. That's really important for crews fixing potholes, and potholes are a universal problem.

There's a tremendous amount of byproduct from iron mining in my region alone. Millions of tons annually. Big rocks for armor stone along shorelines. Crushed stone for railroad ballast.

Another project focused on magnetite, a very good microwave absorber. I was doing a presentation about byproducts, and a gentleman approached me afterward about using mobile microwaves to de-ice roads.

We started a project with regular kitchen microwaves and samples. He had been dabbling in microwaves to thaw frozen ground in winter to access buried utilities. In the past, they would bring a heater with propane tanks, put a tent over the ground, let it cook for a couple of days. He developed a truck-mounted microwave that could thaw the ground in a couple of hours.

We thought this might work for pothole repair. Using microwave technology and byproducts containing magnetite, you could repair potholes in subzero weather. We're also working with iron mining tailings, grainy like sand with good friction characteristics. Being used as friction coating on bridge decks and roadways to prevent skidding.

Having a hand in those two products, the pothole repair and friction coating, it's pretty neat. Seeing materials used in tangible ways that contribute to safety. It's gratifying. The tough part: customers can be state departments of transportation, and they're slow to embrace new technologies. Also, the microwave only works on asphalt. If you microwave concrete, it explodes.

The goal is to deploy this at a much larger scale: a microwave system that can replace traditional asphalt plants. Asphalt plants use a lot of fossil fuel and create fumes. That's why you see them on city perimeters.

Microwave technology can heat asphalt and reduce emissions. You could set up a plant close to your job site. You can apply the same technology to biosolids and treat them with microwaves to kill pathogens so they can be used as a soil amendment.

You may be imagining a kitchen microwave at 1 kilowatt. Imagine several 100-kilowatt units.

A lot of my knowledge has been gained by osmosis, working with different people. You have to be a good listener. People have stories to tell. When you meet someone with a story, sit and listen. If you want knowledge to continue, you need to engage people.

The thing I'm proudest of is having such a network of people I could draw on over the years. I may not be the best person to do all the research, but I can find someone who would be. I work in mining, and a lot of people don't like mining because it's disruptive. Landscapes are changed. It's left a negative environmental legacy.

Part of my issue is I know that nothing is risk free or has zero impact. Zero only exists in math.

I know our energy future will be reliant on minerals. There's a balance. What people don't know is that despite the negative track record, the industry is changing. Companies realize they won't be successful if they mess up.

We need to get that conversation going so people aren't talking past each other. I've sat in on public hearings for mining, and I don't want to say it turns into a food fight, but it's pretty close. All sides cherry pick data. They don't engage directly.

If we can engage realistically, the future is quite bright. The debate itself is holding us back. Society has been thinking about these issues since the Carter administration. It's frustrating when I think about the progress we could have made if we had given environmental issues the moral equivalent of a war. We could be so much further.

It makes me afraid of the future because I'm not sure how much progress we can make under current circumstances. The advice I would give: read, read, read. Ask questions. Be curious. Learn how to distinguish what's reliable and what's crap.

Look for opportunities to say yes, to take new paths. Progress isn't a straight line. Exposing yourself to people smarter than you is stimulating. It gets you out of your comfort zone.

That's how we make progress.

The Energy Data Scientist

Texas

There was a rail disaster in Quebec that happened because the company didn't know the vapor pressure accurately enough. They put it in the wrong container. When it rolled down the hill, it exploded.

All these little measurements matter because this stuff is dangerous. There are specifications you have to hit.

That's the kind of work I do now. I'm a data scientist at an oil and gas consultancy. When I started, there were 12 of us. Now there are over a hundred. We're moving further out of oil and gas. I would say we're more of a carbon company right now. I have a PhD in pure math. I realized there weren't jobs in academia, so I went into finance. It was terrible. This is better.

Mostly I work on research and development, trying to come up with frameworks to answer questions.

For instance, you have a gas plant, and the amount of stuff that comes in is not the same as what goes out. What happened? Where did it go? I'd create a mathematical framework to handle it. Then we'll know if it's your meter, if sample error, or if the gas was vented or flared.

I also help people understand how frequently they should sample their oil and gas. When oil arrives, I'll take a sample and assume the quality is the same for the entire batch. But the density or vapor pressure or sulfur might fluctuate.

That's bad.

High vapor pressure means corroding pipes, which leads to explosions. People come to me asking how often they should sample. A mathematical framework can help them know if their light and heavy oil are going to stratify in the tank.

A. Kelly, *Good Work*, Women in Engineering and Science,
https://doi.org/10.1007/978-3-032-27547-9_11

I get to do a bunch of science and math to figure this out. One exciting project is this group of PhDs who have engineered normal crop and wetland plants to have more roots, roots with a higher capacity for sequestration. They can pull carbon out of the air and keep it in a way that isn't going to biodegrade quickly. The root structures are good for rebuilding wetlands.

That's the kind of work I'm excited about now. The president of our company is terrifyingly intelligent. I'm saying this as someone with a PhD in math. That doesn't mean I'm intelligent. It just means I've thought through problems deeply. But our president thinks through things 10 times faster and deeper than most people ever could.

We recently had a client asking about methane emissions. They had sensors but wanted to know what their real emissions were if their sensors were working. The sales team came to me to go through the data. One service we offer is a digital twin. We take the oil and gas meters, flow meters, and tank levels. These meters have properties that are hard to measure. You have to pull a sample in a pressurized container and do a bunch of stuff. It's a pain in the butt.

The digital twin uses real-time data and baseline testing and builds a model using petrochemical formulas and machine learning. It tells them, at all times, what the vapor pressure is in their lines.

It can save hundreds of thousands of dollars because clients know their product is within specifications.

These are difficult, deep models. And it's neat. Sometimes I'm doing something dramatically different every week. I build initial code for proof of concept. Sometimes I get to plan really interesting stuff. Sometimes someone asks a question and I go read about it and try to figure it out.

I would love a little less breakneck pace. Research and development should not be done at a breakneck pace for no reason. I really like talking to clients. On one project, the client would ask questions, and I got to look through the data and make visualizations. It was delightful.

Graphs are great because they let people internalize data easily, as long as it's presented clearly. Customers think you're their new best friend if you do a good job.

The advice I give: assume a VP of strategy is reading your work, and they're only going to read one page. I feel good about my work. Gas companies have information now that allows them to make informed decisions. If everyone used our system, the probability of disasters like that rail explosion would be much lower.

We help gas companies reduce emissions by figuring out where leaks are coming from. We help them find problems they might not have detected. Then we can fix them.

When I first started in this industry, I didn't get the sense that most companies cared about reducing emissions. That's changing.

That makes me feel heartened.

The plant project, that's one of the areas I'm excited to work in. It's not just reduction of emissions; it's removal of emissions.

There are incredibly smart people working on these problems even though you don't hear about them.

It makes me feel more hopeful than I did before I started working for this company.

The Climate Science Student

Pacific Northwest

I am a PhD candidate studying environmental microbiology. I work at a public land-grant university, which means we're funded through the state and have a mission to serve the region, specifically studying agriculture, oceanography, and geology.

I grew up on a farm and always loved the environment. I loved seeing seasonal cycles of leaves falling and regrowing, watching water levels of creeks falling and rising. I've always felt connected to nature.

I came from a very rural family and community, but I was always encouraged to go to college. I wanted to do science and everyone told me I could do it. I study glacial melt by collecting samples from permafrost ponds. To get to the ponds, where it's swampy and muddy, we use an airboat. When I saw the airboat for the first time, I was absolutely thrilled. I was so excited, but really nervous.

I put on this big motorcycle helmet and realized we were really going to ride on this boat. It was really loud, and I'm afraid of swimming.

Once we were on, we buzzed around the delta near the Gulf of Alaska over open water. It was just fantastic. I had a blast. I think I was laughing and whooping the whole time. It was pure excitement.

But of course we weren't just there for fun. We were there to study Alaskan microbes. Once we got to the ponds, I put on a dry suit. A dry suit is different from a wetsuit. Your skin stays totally dry underneath. It's a full-body suit over your clothes.

Because it's my research, I got to wear the suit. Or rather, I had to.

The ponds are a couple of meters deep. I'm about 5′5″ so the muck, mud, and water came up to my neck.

A. Kelly, *Good Work*, Women in Engineering and Science,
https://doi.org/10.1007/978-3-032-27547-9_12

We shoved this plastic tube into the bottom of the pond, dug sediment out around it, then capped it from both bottom and top before lifting it up. I had to dive down under the water and hold my breath to get the samples.

To sink down, we had to get all the air out of the dry suit. The suit had a little valve on the arm. I opened it and crunched myself into the smallest ball you can imagine to get every last bit of air out. Then when I stood up, the suit was all suction-sealed to me.

Sometimes suits would have a little air left that made my body float. I got better at scrunching into a tiny ball with practice. Once in the pond, I'd have someone push me down to the bottom and hold me there so I wouldn't pop up.

It's not glamorous, but it's a great time. The area we're researching is influenced by glaciers. The climate is warming and glaciers are melting at an increased rate. This is especially true in high-latitude environments like Alaska because they're warming disproportionately.

When glaciers melt, they release all their metals and nutrients and organic matter that were trapped in ice into downstream systems. Those systems are changing, and we want to characterize that change.

We believe glaciers and the matter they release can actually change how systems process carbon, which impacts climate change.

It's widely believed that a system like melting glaciers would produce and release methane into the atmosphere. But that's not what we think is happening.

We think the melting permafrost is activating microbes that were frozen, and those microbes can then consume methane. In other words, as glaciers melt, they may be allowing microbes to consume methane.

It would be such great news. Science like this can take long hours in the lab. My family noticed I was spending a lot of time focused on research.

My little sister is a political activist. She approached me and told me I was getting too far into my science wormhole. I was turning into a science gremlin.

She knew I was doing climate science, and it was important. She made it her personal mission.

One of our representatives was coming to a climate rally, and my sister made me get up on stage and speak about climate impacts.

I received training in advance from my sister's staff. They helped me write the speech and practiced it with me. In the shower the night after training, I gave the speech to myself like 15 times in a row. I pumped myself up and didn't sleep that night. The next morning I repeated the speech my entire drive to the rally.

I was so nervous. So jittery. I was the second speaker. I couldn't tell you what the first speaker said because I was trying not to pass out.

When I heard my name called, I got up there and gave my speech exactly how I practiced. I couldn't even hear myself talk, but I knew it was right.

I listed what we're seeing as climate impacts: Atlantic currents destabilizing, the Amazon rainforest becoming a carbon source rather than a carbon sink.

It was so exhilarating.

My little sister was so proud she was jumping up and down. I had people come up afterward and ask questions. Thanks to my sister's preparations, I was ready with citations and literature.

Then we got chicken and waffles. That day got me thinking about what else I could do. The people who come to these events are stressed and worried and want to know more.

Climate change is this huge existential issue that others and I grapple with every day. It's never enough. That's my real driver getting myself into the position I'm in now.

I was able to see the climate changing even when I was younger, and I want to understand these changes. I wish within my heart of hearts that climate change wasn't happening.

I know I will spend the rest of my life putting all my effort into researching carbon. It's really important.

Seeing my little sister in politics shows me that real change comes from the Senate and the federal government. I can put forth all this research, but if it doesn't reach politicians, it may not make a difference. Since then, she has been making me talk to politicians at local campaign events. Last summer, I was a concerned citizen scientist on a panel for a local environmental group.

I say she makes me, but I really enjoy it. That's why I go to rallies and seminars and conduct webinars. I've been writing letters.

I know it's the third time I've said this, but it just feels like it's never enough.

At least in my area and my generation, we have strong grassroots organizations. Everyone can see this affects not only us but the entire world and future generations. My sister inspires me. She taught me that if you want to talk to someone, you should just write them. You can write the head of the EPA. You can write senators from states that aren't even yours.

Reach out to all the people you normally wouldn't feel supported by. They're just people. They're just humans.

I used to have them up on this pedestal where they were unreachable superhumans. Then I sent a letter, and the head of the EPA wrote me back. He responded.

His letter is hanging up on my refrigerator.

It was so encouraging. I could see all these organizations fighting tooth and nail, and it felt amazing to get confirmation that he was also fighting.

Maybe all these rallies and things can really make a difference. Maybe our work does reach the capital where real change is made.

The Behavior Researcher

Canada

We're designing a game to teach people about energy. You start with a building, you have a carbon footprint, and then you start making changes. You change out the windows, you change the insulation, but you have a budget. Every choice has tradeoffs.

It's part of a bigger question we're trying to answer: how do you get people to actually care about their energy use? I'm a PhD student at a university, working in a research lab that studies how people interact with buildings. Not the engineering side, the human side. How do they use their thermostats? What do they understand and what they don't? What makes them drop out of an energy savings program?

My focus is on demand response, when utilities need to reduce load during peak hours. We model how many people will actually adjust their thermostats when asked, what temperatures they'll tolerate, and what makes them give up and override the system.

In Montreal, where I am, this mostly happens in winter for heating. The main challenge is figuring out how to make these programs work for people, not against them. Right now, I'm developing a survey to understand demand response better. But we're making it differently from normal surveys. It's scenario-based. We aren't just asking what time you do laundry, but presenting different situations and trying to identify what people prefer.

We're also trying to gamify the survey. That's where the building game comes in. When I finish the survey, I'll collaborate with the gamification team. It's still in the idea phase, but I hope it will encourage people to be more engaged. I've also been using data from a government time-of-use survey. This shows how people spend their time during the day. I can take data for a city or district and relate it to people's

A. Kelly, *Good Work*, Women in Engineering and Science,
https://doi.org/10.1007/978-3-032-27547-9_13

characteristics. For example, a 25–35-year-old male, I can put some characteristics on him like he's leaving his home and he's working at this time. Different customer profiles emerge. We have a big team. I don't know how to describe the structure. We are a big institute in various cities, and there are several specific labs under the institute housed in the university. We have a transportation team, a buildings team, and a gamification team. My team is on occupants.

Because I'm a PhD student, I have an advisor and I work with him directly. There are five PhD students in the lab. Sometimes we work on our own project because a PhD is individual work. Sometimes we have a project and help each other. I'm good with programming and coding. If someone wants to code something, I'll do it. Then there's another with an architecture background that knows a lot about buildings and envelopes. We calibrate based on skills on a case-by-case basis. I studied in Egypt for my bachelor's and master's.. When I finished my bachelor's degree, I didn't know anything about sustainability or environment. These topics are not really in the daily talk of the field in Egypt. I started like a typical mechanical engineer that was not focused on buildings.

When I finished my studies, I worked in a normal company as an HVAC engineer. But I liked research more. There was nothing new. You were repeating the same thing. So I went to do a master's degree.

I worked more in energy storage and found that I liked research. But I got mixed feelings about academia. I don't like academia. I like research. I want to apply it and work in industry. When I finished my master's, I wanted to work in a field closer to industry. Building energy was an interesting topic. A lot of industries and governments are putting money into retrofitting or reaching zero-net energy buildings. So I shifted from mechanical engineering to building engineering. I started 2 years ago, so I'm finished with classes and in the second half of my program. I don't have any regrets, but I worry. I'm not sure if I will find opportunities in industry. Sometimes you feel okay; I do a lot of these things in a PhD, but how will I land that in a company? I think I'm not the only one. A lot of my colleagues are in a constant state of worry. You do a PhD, you do a lot of work, and then what? If I told you I had a plan, I'd be lying.

You try as much as possible to learn and work in different projects during your PhD so you have a good perspective about the spectrum and what can be done. My plan is to work in industry, not academia. I like academia because I get to solve problems from scratch. I think that even if you start from scratch, even if you don't know anything, you will find someone who has done something like what you're doing. You will never find the exact same problem, but you'll find a piece of it. You start to get ideas bit by bit. You suddenly find that you completed the puzzle. While you are working on it, you'll never know.

Like in the first stage of my research with the time of use data. I started with no clue. What is this? Where do I start? It's a big and difficult project. But I just start with any step. And step by step I complete the picture. This is my approach, generally. Things change with time. I like to be optimistic. I feel like we can combat climate change, but not doing business like we do now. I hope some things will change. My idea is that things will change when it starts to directly impact people.

Electric Vehicles are not a new thing, but when they start to be a business and make a profit and not make a car that looks like a car from a sci-fi movie, then people will start to like them, and the business will boom. A lot of things in the climate change field are really good when you see the research. Things are working efficiently. We just need people to get it and convert it into a business that generates a profit.

When you speak with people, don't just give theoretical speeches without looking at their daily lives. You cannot just go to people working in an oil field and tell them what they don't know. I don't have dreams of changing the world. I'm more pragmatic than that. But at the same time, I have to have an impact. I want to work in industry or in a company that has as positive an impact as possible. I don't want to work for a construction company where I just end up making energy models. I don't want to have the minimum impact. I don't want to work for an industry that is only for profit.

I could work for a consultancy. A consultant is working for profit, but they have a wide range of impacts. They consult for the government, for different companies, so they are able to make a change. I could also work in a national lab creating standards for the government.

I'm interested in fields that look toward the future. I'm not sure how it goes, but I'll figure it out. Sometimes maybe you have a big impact, but you don't know it. I know some professors that taught me courses in university that have impacted my life. People can have impacts that won't even be recognized in your lifetime.

The Energy Software Developer

Pacific Northwest

I work for a company that is largely an energy efficiency implementation firm. We use data and software to deliver scalable energy efficiency programs and products that focus on small- and medium-sized businesses in the United States and a little bit in Canada. The mission is to have a positive impact on climate change. To make a legitimate impact and to do it in a way that is actually impactful.

There's an interest in not just doing things but in only doing things that will work. We have a pay for performance mandate, so we only get paid for what we achieve and what can be verified by third-party evaluators. The climate change impact has to be measurable.

We implement energy efficiency projects. We calculate savings, but we only get paid on the savings that are calculated by a third-party evaluator. We have a fee structure where we get paid based on the number of kilowatt-hours that we save. I am responsible for leading the technical component of our measurement of energy savings and the software for design and implementation. I also support other members of my team, the measurement and verification team. I participate in industry working groups to share knowledge and to move the standards for how we measure things forward, to make measurement more accessible, more reliable, and more effective at combating climate change.

I also have a broader data science and analytics role where I lead the technical direction on projects. I help identify opportunities for energy savings algorithmically and evaluate the performance of our program. I work on machine learning and other methodologies. I lead the data infrastructure that supports all that work. I have a pure math background. When I say pure math, I mean math for the sake of math. It's just proving things. It's unlikely to ever have a practical impact on the world.

A. Kelly, *Good Work*, Women in Engineering and Science,
https://doi.org/10.1007/978-3-032-27547-9_14

But then, when I was in a math PhD program, I got involved in image analysis for medical imaging, basically figuring out how to read x-rays and MRIs without a doctor, using math to automatically detect anomalies. As the years went on, the approach changed. People stopped caring about the math approaches and got more interested in machine learning. I started learning and realized there were more jobs in the machine learning space.

To use machine learning, you have some data about anything, and some of the data can be related in one way or another. You try to figure out the relationship between different data points. If you're doing physics, you derive an equation and ask if the data fits. Machine learning tells a computer to figure out what the equation is. There's a lot of black magic and heuristics. The rule of thumb is that it's a big black box you throw data into. I got a couple of contracts to do machine learning work and left my PhD program. One of those contracts was with my current company, and it was a really good fit, from the people to the mission. That was a little over 4 years ago. I got with the mission to try to use data science to improve the things we could do. My role grew into the measurement and verification space and the analytics team.

In the beginning, I was just learning the technical skills, and the hard part was getting my code to run. But eventually, I got better at coding, and the hard part changed to making sure I know what the problem is that I'm trying to solve. Do I have the right kinds of data? Do I have enough of it? Is the quality good enough? Is the algorithm working? Here's an example. A classic computer vision problem is pictures of cats and dogs, and you want an algorithm that can look at a picture it's never seen before and tell you if it's a cat or a dog. You use your algorithm, and you get 80% accuracy.

When you're working for a company, that might not be good enough. Back when the problem was cats and dogs, it didn't really impact anything to be wrong. But in industry, if you say it's a dog and you're wrong: that's a million dollars. You have to be careful when working with machine learning.

Different people have different visions for how machine learning can be used. I have to understand all the different people who want to use what you're building and exactly what they want to use it for. I get pedantically hung up on making sure that people understand how to use what I've made. I don't want to deliver something people are going to misuse. The people I build things for don't know what they are going to learn from the information and that's a problem too. I have to uncover what they want to learn through conversation and by digging into the context and details. Then I have to write a lot about it. I didn't think I had a writing job, but it's all I've done in the last month.

When the company was smaller, I didn't have to write much. I would just make things and it was fine. But now I don't control all of the pieces and there are more stakeholders. If I want to do a thing, now I need to write a proposal. Then we do it and I need to write about the results. There's more documentation and that's good. Our company has quadrupled in size since I started, from 20 to over 80 people. When things are small, it's more agile. You know everybody and you just have to be scrappy all the time. You're not going to get a whole bunch of new resources. You just have to work with what you've got.

I didn't have enough experience to know what tools you can buy or who you could hire. It's not intuitive to me to solve problems by getting new tools. I'm just going to try to figure it out because that's what I do. Scrappy.

Also, when things are small, you're working with the final decision-makers. As the company got bigger, it got more structured, and I lost access points to the decision-makers. My team is small. I'm a machine learning engineer, and I manage a data analyst (who is great), and we are hiring another analyst now. I have a lot to keep track of.

At the end of every day, I write down the things that I want to make progress on. It's a rolling to-do list. In the morning, I check my email and Slack and update the things I want to get done. The advice is to do your creative work first thing in the morning. Otherwise, I just choose the most fun option from my to-do list. I might have writing, scoping projects, researching, developing, or data science. I pick the thing that seems fun. Like data science. I think the company is having a really positive impact. We're making a positive impact. I feel good. There are lots of data science and development jobs that are not benefiting society, let alone are actively harmful. It's nice to do something where I'm not hurting anybody. I'm helping some people.

I think I've done some work that has been impactful, particularly in the measurement and verification space. That directly enables our program to exist. Other people are still doing the real work of calling people and combating climate change. I'm in a supporting role. The data analytics work is one more step abstracted from that. I'm working on saving energy, but not doing it directly. It's abstract.

I enable other people on my team to work more efficiently or do something they can't do without machine learning. Then people downstream can use my things and it'll have a positive impact.

I just like solving cool problems in elegant ways. That's really who I am at the core of my being.

Part III
Work With People and Policy

I need architecture, not movements.

That sounds cold, maybe. But that's not what I mean. I need a constant driving pressure because the work is getting done. Not people screaming in the streets. I need systems that function without anyone having to be a hero.

The people in these eight chapters don't build anything physical. They don't run regressions. They move people. Is that harder?

They also lose. One of them spent 2 years passing a bill to ban Styrofoam cups and takeout containers. They passed student government resolutions. They were in the news for river cleanups. They collected thousands of petitions. It was voted on in the House on Earth Day. Then the fire alarm went off. Everyone had to leave. The vote got pushed. It failed in the Senate that summer.

But even after a tough loss, there's always another campaign. The same bill that failed was passed a few years later. It takes time. Sometimes if you haven't talked to enough people, you keep organizing.

One of them builds shade equity coalitions because urban trees aren't a nice-to-have, they're essential for human life. During the heat dome, almost everyone who died lived in neighborhoods with low shade canopy.

One of them works on federal energy policy, thinking about workforce pipelines and first-generation college students. That person is the fed. They're thinking holistically in ways most people never have to.

The EV policy designer says our kids are not going to know how to pump gas. That's moving faster than anyone expected.

That's why I know I need millions of people and not thousands.

These chapters are about what happens when one person decides to help other people see their place in the system. It doesn't make for great photos. But I think it's probably how change actually happens.

The Campus Organizer

Pacific Northwest

I'm the campus organizer. I teach students how to be activists and help them run the campaigns they care about.

One week, I might be recruiting freshmen who care about climate change. I could be working on funding or hosting a lobby day. Every day is different. But my top goal is giving students the skills they need.

To be an activist, you have to be willing to step outside your comfort zone. Activism is about asking people to take action: volunteering, signing a petition, running an entire campaign. You have to be prepared to confidently ask someone to do something.

The top skills are managing your time and public speaking. I started when I was a student volunteer. We would do announcements in classes, asking if people wanted to get involved.

The first time I did an announcement, it was so terrifying. I blacked out. I don't even remember it.

I worked with my campus organizer. She had me practice, gave me tips, shadowed me, and gave feedback. I've done dozens since. The more you do it, the more comfortable you get. It is shockingly hard to get people involved. Not because they aren't interested. Imagine you're scrolling on Instagram and see an announcement for a climate event tonight. You'll think, "Oh, that sounds cool. I should go." But you aren't going to go.

We do giant recruitment drives at the start of each term: tables, class announcements, club meetings, fairs. Have them fill out an interest form; then we call them. Every single night we have phone banks where you can talk one-on-one.

A. Kelly, *Good Work*, Women in Engineering and Science,
https://doi.org/10.1007/978-3-032-27547-9_15

Once you have that connection, you can say, "Hey, want to come to this meeting tomorrow?" Of course, only half the people who say yes actually come. We work on public interest issues, things the majority already agrees on. Like not using single-use plastic. But there's a special interest with a lot of power and influence standing in the way. Fossil fuel industries, chemical industries. Industries invested in the way things are.

The campaigns typically involve climate change: passing a clean cars bill, getting commitments to renewable energy. We helped ban plastic grocery bags and foam containers in one state. We got another state to commit to a hundred percent clean electricity.

How you win is by mobilizing people. If you're an elected official, what do you care about? Getting re-elected. If you see huge numbers of constituents saying "Pass this bill," that sticks out. When you lose, it's hard. People will tell you they'll vote "yes" to your face, then vote against clean energy the very next day.

My most heartbreaking loss was my first big loss. I was young and naive and thought, "Well, it's enough to be right. It's enough that I care."

We were passing a bill banning Styrofoam cups and takeout containers. We'd worked on it for 2 years. We'd passed student government resolutions. We were in the news for river cleanups. We collected thousands of petitions. We canvassed door to door. It was voted on in the House on Earth Day.

We were sure it would work. No one would vote "no" on Earth Day.

Then the fire alarm went off. Everyone had to leave. The vote got pushed.

We held emergency phone banks that night. We called everyone we knew, asking them to call their elected officials. It passed the House by the skin of our teeth.

Then it failed in the Senate that summer.

People were just crying. You don't always win. But even after a tough loss, there's always another campaign. We passed a healthcare bill recently. Students were absolutely convinced it would pass. They didn't know the very same bill had failed a few years earlier.

It takes time. One state took 8 years to ban plastic grocery bags. Sometimes you haven't talked to enough people, so you keep organizing.

Even winning doesn't always work. We banned single-use plastic grocery bags, but the plastics industry had a workaround ready. They made thicker bags relabeled as "reusable," even though everyone still used them once. So messed up. College students are uniquely positioned to be activists. You're passionate and engaged. You have more free time. You can find an hour between classes to make phone calls.

My first lobby day was crazy. I had never lobbied before, and I was the staff person. It wasn't until that week that I knew you could just walk into an office and meet with your elected officials.

We scheduled a ton of meetings. Students from across the state broke into different groups. I was leading a group, preparing them to speak, running around the Capitol finding rooms. I was nervous, but I hid it.

In person you might just run into the senator. When I was a student, we ran into one leaving a committee room to go to the bathroom. We literally lobbied him in the hallway.

You can't do that over Zoom. I feel great most of the time. Even when we don't win campaigns, I feel great about the impact I have. Even if students aren't going to be full-time activists, they're going to do incredible things.

I've seen students come in really nervous and leave as confident public speakers. I knew them when they were shy first-year students.

It's people power. Look at us, just college students.

Not even 5 years ago, I would just use Styrofoam and not think about it. Now states have banned it and the whole West Coast has committed to clean energy. Oh my God. I feel so many things.

I want people to get involved, even if it's scary. Everyone else doing it also thinks it's scary. They're just pretending it's not.

Do what you can. You aren't going to do everything on your own. But with others, you can make a difference. If you join a group of 30 people, and each of those 3 reaches out to 10 people…

That is what makes a difference.

The EV Policy Specialist

Pacific Northwest

Our kids are not going to know how to pump gas.

Car manufacturers are saying they'll be making 100% EVs by 2030. It's going to be widespread way quicker than people realize. A lot of time policy is really lagging on climate issues, so the fact that this is happening more quickly than we expected is rare. And awesome. My company owns and operates one of the nation's largest public fast-charging networks for EVs.

My role is market development and public policy. I work with public utilities commissions to advocate for EV charging. I work at legislatures improving laws that impact our company. I work directly with municipalities on different issues.

At any given time, I might be working on multiple regulatory proceedings at different stages. Lots of writing, lots of internal meetings, and strategy work. But the biggest focus is external stakeholder meetings. EV charging has a lot of different stakeholders. State agencies that provide public funding. Energy offices, environmental quality departments. Departments of transportation are becoming stakeholders now that they're getting federal funds for infrastructure.

It's surprising, but transportation departments hadn't been involved until recently. There's a learning curve. A lot of time they just haven't done programs like this before.

The real lesson is they need to work with other agencies. Not reinvent the wheel. Infrastructure is supposed to be an interagency effort. I work with public utility commissions in their regulatory processes. It starts with a utility filing something like a transportation electrification plan. We become an interested party and intervene because we have a tangible interest.

A. Kelly, *Good Work*, Women in Engineering and Science,
https://doi.org/10.1007/978-3-032-27547-9_16

Once we're a party, we provide testimony. There's an evidentiary hearing where we cross-examine different witnesses. Then the commission makes a decision. It's usually a 6- or 8-month process.

There's a lot of work. Research and writing, but also outreach and building coalitions. We need to understand where everyone is and find common interests.

Then there are settlements. Sometimes a partial settlement resolves only a few issues. It takes a lot of meetings. Then it has to be written up, and everyone has to agree to every word. If you can't settle, you'll have to fight about it in front of a judge.

I always thought regulatory and policy affairs sounded boring, but it's actually really exciting. The formality and permanence of it all make it extra exciting. To start conversations with agencies, sometimes I'll just reach out to a random person at a department of transportation trying to find the right contact. Some are really lost. Some are really far ahead. Every conversation is different.

We're trying to provide best practices. We've compiled our experiences from all over the country. We've seen good programs, but we've also been part of poorly run programs. We want to warn folks there's a right way and wrong way to do things.

One good example: applications for a program. You don't want first-come, first-served. You want competitive applications. First-come, first-served means you get a flood of varying quality from whoever clicked fastest. A lot of those projects won't go through. We recommend solicitations with a scoring rubric. It's hard to place chargers. There's cost-effectiveness, power availability, and distance to highways. We don't recommend agencies be rigid. If you're one mile from a highway, great. Five miles? A few less points, but still okay.

Recently, we've started working with municipalities on permitting timelines. It was the same thing that used to happen with solar. I'll talk to local folks at the permitting department and figure out how to streamline things.

There are tweaks within each department that can make or break a project. Maybe we need the city code changed. Maybe we need legislation. Maybe we need additional funding. All of these pieces are tied together. We need to move the needle forward on climate change. There are barriers in every industry, and it's my job to figure out what's slowing us down, because we are ready to go.

I love finding institutions like local governments and changing their long-held processes. You have to meet folks where they are.

Across the country, we're seeing a lot more support for transportation electrification. We don't have to argue that we need it anymore. The federal infrastructure bill gave credence to the cause. Now we're past talking about *if* we should do it. We've moved on to *how*.

If I talk to a municipality with no interest in climate change, we go the economic development route, bringing business to the city. I feel good about the impact I'm having. Before, I was working in a wider variety of areas for state policy, higher level, more removed than what I'm doing now.

Transportation electrification is blowing up. It's a huge thing right now. I feel like I've gotten in at a good time.

The Case Study Curator

Canada

I manage a website that shares success stories from climate change initiatives. I've been doing this for 28 years. I suppose you could say I'm a professional storyteller, except the stories I tell are about success.

My background is in social marketing, which sounds like it's about Facebook, but it's not. It's about using marketing principles to change behavior for social good. Instead of selling soap, you're selling the idea that people should walk more, or use less energy, or weathercasters should talk about climate change on TV.

Since I began, the site has been mostly about climate change. Communicating this information is the one thing I seem to be doing that no one else is. It's where I make my contribution. I focus on curating case studies on building energy use and sustainable transportation. I select them and get them peer-reviewed. The case studies get rated by panelists. We designate the best ones as landmark case studies.

It's vital that the impacts are measurable. All 150-plus case studies have measurable impacts. We look at how they piloted it and assessed it. We ask how people addressed their barriers. We want to know how they're financed, how they measure impact. A recent landmark case study is a program for TV weathercasters. It trains American weathercasters to report on local impacts of climate change. It helps combat the narrative that climate change doesn't impact people locally.

The program provides participating newscasters with weekly story packages: local data and broadcast-quality graphics. They engaged conflict mediators who work with opinion-leading weathercasters to surface and work through entrenched conflicts.

After just a few years of the program, there were approximately a thousand participating weathercasters in almost 500 TV stations. They showed that viewers

A. Kelly, *Good Work*, Women in Engineering and Science, https://doi.org/10.1007/978-3-032-27547-9_17

exposed to climate education from weathercasters understand that climate change is already impacting us here and now. What I do depends on the time of year. I'm starting a new cycle, and the deadline is 4 months away. Between now and then I'm looking for interesting stories and research from peer-reviewed journals.

I focus on climate change adaptation and mitigation, energy, transportation, and waste. I skim and file articles, then do the deep dive later. I keep track by only logging the good ones with measured impacts.

I also put out a call, and people can submit their work. The panels comprise people I've known and respected throughout my career. I try to get a good balance from the United States and Canada, with a little global representation. I try to include consultants, academics, and practitioners.

I usually have about eight people in each panel. I put all the case studies in a package with scoring sheets. Reviewers ask if they agree with the methods and the identified risks. They make comments and fill out scoring packets.

When scores come back, hopefully we'll have one with the best score. I'll contact them and let them know they've been designated a landmark case study.

The promotion involves posting the case study around the web: LinkedIn groups, my userbase, listservs. I know people use the case studies as starting points when planning projects. That's happening around the world. Some instructors use it for their classes. Some people use it as a planning tool. Social marketing has changed a lot. There are two parents: marketing and social psychology. Marketing gives us market research and measuring results. Social psychology gives us how people think and change.

The product is the behavior change we want. Let's say we want people to walk more and not use single-occupancy vehicles. That's the product.

One case study was from a cement company doing carbon sequestration. The cement industry has the second highest carbon dioxide emissions of any industry. The product was cement that recaptures CO_2 from the atmosphere. It's now in more than 300 plants worldwide. I published it so people could find it and learn from it. Stories like these give me hope. I'm seeing more companies take real action on climate change. I'm hopeful we're seeing more consensus that this is human-caused and that we'd better do something about it.

It does dismay me to see distractions. We need to focus all our energy on climate change. There are lots of people resigned to human extinction, and they dismay me.

But the work continues. Finding measurable success in a challenging world.

The Federal Energy Official

Washington DC

I'm the fed.

I work as a researcher and program manager for an advanced energy consortium at a national laboratory. I wear many hats. In all the work we do, that's who I represent. That's why I say I'm the fed.

I act as a broker between big energy projects. If they involve nuclear energy or hydrogen storage, they partner with us. Additionally, I'm involved in research involving reactor technology across the country.

Regionally, I focus on workforce development, creating a talent pipeline. Building the nation's energy workforce is a big thing. There's concern about being able to train enough people. We're trying to plan for it. I work on student programs with business schools and schools of public service. I assist programs with underrepresented students in STEM leading to energy fields.

Underrepresented students may not have access to the same kinds of knowledge as others. I was a first-generation student. Because my parents didn't go to college, I didn't know about half of the resources I could have used. One program I worked with created a first research experience for students. I helped bring students to the lab to work for 10 weeks with a professional researcher. The programs are aimed at first-generation college students. They live together and go to work every day on a real science project.

We need to help first-generation students the most. They often don't know what kinds of internships exist and what skills they need.

Getting my first job out of school was awful because I didn't really care about networking. I don't enjoy schmoozing.

A. Kelly, *Good Work*, Women in Engineering and Science,
https://doi.org/10.1007/978-3-032-27547-9_18

I don't think most autistic people, like me, enjoy small talk. It's really, really challenging.

We show them what exists and introduce them to people they wouldn't otherwise get to know. Even if students leave the lab only knowing me, I can route them later. Having access to me is having access to my network. I also work on proposals.

There are federal funding opportunities all the time. Scientists need to write proposals to get money for their work. But applicants now need to include components about diversity.

We are asking nerds trained in engineering and hard science to write proposals that include diversity, equity, inclusion, and environmental justice. They don't know how to do it.

A big part of my day-to-day is putting together plans for the scientists. I write the justice parts of proposals because we don't have those skills in the marketplace yet. The folks with expertise in nuclear technology don't also have expertise in social science.

Even though I still feel like a little kid, I put together the webinar for all 5000 people at the lab on how to align on the importance of diversity. I help people list out what they already know. Work doesn't exist in a vacuum. The reason scientists do research to improve energy is because there are people who consume energy.

I think that gets lost. Lost in the noise.

I guide them through learning how to say what their work means, who their end user is, and who will benefit. I teach them to zoom out and see how they fit into the greater context.

These scientists have a hard time unzipping the nerd brain and realizing they work on energy research because people need energy to do things.

They'll get lost and focus on diversity for diversity's sake. They got in the habit of saying "Our principal investigator is a woman." They can't rely on that anymore. When I see that in a proposal, I sit them down and ask how having a woman leading the project is impactful.

It's like when your grandpa tells a joke and you ask "Why is that funny?"

Sometimes it's hard for them to focus on who is really benefiting. Teaching them sometimes makes me want to rip my face off. People have a really hard time seeing the big picture of the science they do and the impact it has on communities. I walk people through identifying what communities they're helping.

I find talking to scientists about their work really interesting. I don't need them to get it right the first time. I just need them to put something down. I'll ask them to write down their work and they'll give me a bunch of equations.

I don't know what this math is. It's lovely, but I find myself asking, can you, for the love of God, help me understand what it's used for?

I bridge the gap between whatever is going on in science land and the end users.

Saying that out loud makes it seem really daunting. Now I feel sick. I help them understand we care about societal impacts because any new infrastructure affects communities. We have to build things for our energy future, and building always has impacts.

If you're going to cause a problem, you need to be prepared to solve it. You can't guess the solution. You need communities involved to tell you what they think the problem will be and what solution they would want.

The community doesn't know the ins and outs of the energy transition, but they know there's going to be a big construction project.

Maybe their biggest priority is everyone in town has enough to eat. People want a benefit for giving out their land and labor. Scientists need to figure out how to approach that. I broker these exchanges. One partnership we have is with a tribal entity. The lab is built on their designated land. Their historic lands, their sacred lands.

Our agreement to use the land isn't a greenlight to include them on every project. It means we will always check if they want to be involved because they're interested in preserving their resources.

Whether researchers want to admit it or not, the communities they work with have been negatively impacted by previous projects, reducing access to clean water and damaging air quality. Talking to people about climate change can be tough. I'm in a very conservative state. There's a good chance that when I'm appealing to someone about climate change, they don't think it should be part of their work.

Maybe that's true for them. But it's part of my work. People working on the energy transition may understand both climate science and how it impacts communities. But at the science factory, climate solutions get separated from people solutions.

If you're looking at a graph of energy use and its impacts on climate, you may not see the people muddling around on the earth being impacted.

I'm always trying to infuse people back into it. Trying to get people to see that everything has an impact on some group.

There's only one answer to the question, "Why does our work exist?"

It's because there are people.

The Forest Climate Manager

Pacific Northwest

People need stories. It's how we learn and share culture.

I just started 6 months ago with this organization after 10 years somewhere else. I work for a climate justice organization, a staff collective, meaning all workers get to have a say in what they do. My role is forest climate manager. Before I came on, they didn't have anybody focused on forests.

They did have a fossil fuel resistance team, people working to stop pipelines, to stop oil trains, trying to stop fossil fuel exports. They call it the thin green line. Since this work began 15 or 20 years ago, climate groups trying to stop fossil fuel exports have had amazing success. There actually hasn't been a single fossil fuel terminal built. I got the job the same way I've gotten all my other jobs. I found a cool group and went up to them and told them I thought their work was cool and asked if they wanted to work with me. They said, "Yeah, actually we do."

Then there's the funding piece. They were happy to open the position, but I wasn't in their budget. I told the board I would commit to writing all the grants to fund my position for the first year, and after that they would have to lump me into the budget. I would be indispensable by then.

I've been working on forest conservation for more than 20 years. I know people that fund nonprofits and foundations, so I started putting the word out. I spent the first 2 months saying yes to everything anybody asked me to do. It was amazing to be spacious and say yes, yes, I'll do that. I said yes until I had built enough projects and connections.

My role is to blend together climate groups and forest groups. For the longest time, forest conservation focused on needing to save the trees for the owls, or the salmon, or so we can hike. I was an Earth First! Kid during the 1990s. That's how I

A. Kelly, *Good Work*, Women in Engineering and Science,
https://doi.org/10.1007/978-3-032-27547-9_19

got involved. They sent me to law school, and I became an environmental attorney so I could protect the trees for the critters.

Over time, I learned more about forests, their complexity, and most importantly, their role in sequestering carbon. I've been thinking about why forest groups haven't been talking about climate, and why climate groups haven't been talking about forests. An important project I've worked on was responding to people dying from extreme heat. The idea is that shade is not available in equitable amounts throughout the city. I built a shade equity coalition.

People always think about urban trees as something that increases home values, but they do way more than that. Last year, we had the deadliest natural disaster ever recorded in our state with the heat dome. More people died during it than any other single event.

You know who died? It was almost all people living in neighborhoods with a low shade canopy.

Urban trees aren't a nice-to-have. They're essential for human life. Another important project was collaboratively creating a wildfire messaging guide for climate activists. We learned that climate groups had some messaging incompatible with forest groups around fire. The forest groups knew that fires are natural, and that fear of fire was being used as an excuse to log everything.

Now we have 130 members and 70 different working groups. It's been incredibly useful for the entire region to shift away from silos and share information. We feel like a community. We are building a movement and not feeling despair. We celebrate each other's victories. Right before COVID, there was a National Forest Climate Convergence. A bunch of people from our region couldn't figure out why we never did this at home.

We planned to meet at the largest environmental law conference in the world in March 2020. It got canceled a week before COVID. We pivoted to a listserv. It kept organically growing.

Now we have a whole regional network we couldn't have built if we had kept trying to meet in person. After a few months, we decided to pretend our coalition was real and give it a name. We called it the Pacific Northwest Forest Climate Alliance.

After we published and distributed the fire messaging guide, people who engaged with it did change their messaging. Instead of hearing about backcountry logging, we're hearing more about community preparedness.

Logging isn't the right solution to prevent fires due to climate change. Older forests are more resilient to being burned. Older, more complex forests are going to be better off in most fire conditions. The narrative change is already happening. Five years ago, no one talked about climate change when talking about forests, but now it's commonplace. Having been part of that change feels great.

I want to be part of the values shift from thinking about forests as a commodity to something that's community value. We need to start evaluating things by social cost of carbon and ecosystem services. Intact forests have so much more social and economic value.

We're going to run out of water quickly because of climate change. People say our state will be a great place to weather climate change, but it's not. We're losing our snowpack and glaciers. That's where summer water comes from. Intact forests store a lot of water, and we don't value that. I really enjoy community education and organizing. Movements are great, and very unlike law. In my previous career, I would stop the bad thing with law: stop a pipeline, stop a timber sale. Movements change culture and eliminate the demand for these things in the first place.

People worry about logging jobs. The idea is that if we stopped logging we'd lose jobs. But that's not true. We've already lost jobs due to massive mechanization that started in the 1980s. The industry was able to blame lost jobs on decreased logging due to environmental activism, but if you look at the graphs, timber sales have only been increasing. Production increased because of mechanization, not workforce growth. I feel really good about my work. I left my last job because I kept banging my head against the same brick wall for a decade. It was time to choose a new brick wall. I'm not feeling burnt out because I'm doing new things.

I also believe that taking action on things I believe in is important for my mental health and for creating the world I want to see, even if it's imperfect.

I have two kids old enough to be in their own existential crises around climate change. My daughter is 17 and really worried about what's going to happen. I can't sugar coat it. What I can do is model the fact that we can still choose to act even if the future is unknown.

I'm not hopeful that things are going to be fine. I'm hopeful that humans are really resilient, maybe too resilient. There's a Lord of the Rings quote I keep coming back to:

"It is not our part to master all the tides of the world, but to do what is in us for the succor of those years wherein we are set, uprooting the evil in the fields that we know, so that those who live after may have clean earth to till. What weather they shall have is not ours to rule."

I'm not going to solve it, but I can do what's in front of me. How can I be of service? That's all I can do.

I don't find despair or grief particularly useful emotions. I grieve a lot. You can move through grief. There are days you can't move; then you have to keep going. You don't move through despair. You just sink into it.

I don't think the future needs a bunch of despairing people.

That means people like us are doing what we can in the hopes that there's a future to judge us.

The Utility Partnerships Manager

Southwest

I have two children, and I don't want to look back on these days and feel like I didn't do enough. I want to feel some sense of pride that I made some difference, even if it was incremental, in improving their future and their world.

At the same time, I've done a lot of thinking about how much passion I need to bring to my day. I don't want to burn out. I don't want to become disillusioned and not be my best.

There's some balance one has to strike. These kids are only going to be in my house for another 7 years. If I fully commit to my job and never see them, that's not ideal. If I quit because it's too hard, maybe I won't make as big an impact. There's some balance I'm trying to figure out right now, and I'm not really good at it. I'm behind a screen all day. 100% remote. I probably spend 80% of my time collaborating and 20% on focused, independent work.

My company is a typical utility. I've worked in the public sector, the startup space, and at utilities. Utilities do what they're good at. But doing what you're good at doesn't result in innovation. Doing what you're not good at results in innovation.

There's a lot of momentum. Utilities' thought processes are all about compliance thinking. Given that the public is less and less trusting of corporations, we have a lot of work to do to show up for our customers. There's also tension between those who have been doing this work for decades and those of us who want to do it differently. I'm in the latter category. I've been at the utility for 3 years. Two years ago, I was the municipal customer segment lead, charged with figuring out how to show up for cities with sustainability action plans. They had done all the work of making a plan but needed help executing.

A. Kelly, *Good Work*, Women in Engineering and Science,
https://doi.org/10.1007/978-3-032-27547-9_20

I assembled a team, did economic development to get jobs to the cities, secured grant funding, worked with city account managers. That evolved into a smart grid test bed community. We were going to set up smart grids and embed community outreach to make sure we were engaging with all the right people.

We met challenges because it was a test bed and a bottom-up effort. Not everyone wanted to apply an equity lens to program design. Because of those challenges, we moved that work into the Diversity and Equity Office so we could have more influence and work top-down. My team came from community outreach backgrounds. We hired folks that were at community-based organizations to begin with and had the relationships already. Then we went through the incredibly difficult process of onboarding them to a utility corporation.

There's a big difference between marketing and outreach, and you need to pull that apart.

Community outreach lets you become partners with other organizations and think differently about what partnership looks like. I was able to promote my team and hire someone to replace me, someone with more lived experience and expertise to lead this work, and then I stepped aside. Now I'm a liaison to the state's energy trust organization. I'm a partnership manager. What I do is connect the dots between utility demand response programs and ratepayer-funded energy efficiency programs. I try to sell the peanut butter and jelly sandwich that is demand response plus energy efficiency. And I try to create collaboration.

The collaborations that exist today reduce costs for both organizations and customers, but they're very informal. Historically, we've shown up independently and done what we're good at without thinking about how we complement one another.

The past 20-30 years we've been good at energy efficiency, so we've spent that time investing in it, almost $100 million a year. But energy efficiency has changed. A lot of low-hanging fruit has been picked. Technology has matured, we have climate urgency, and we have equity considerations.

We need to think differently about how to reallocate funds in ways that are more equitable, climate-forward, and technology-aware. We have limited resources. It's not feasible to try to do everything. We have to choose. The impact we'll make may be limited. But if we want to be more targeted, let's get to work.

As the liaison, I ask these questions. I ask my colleagues, staff at the energy trust, and regional organizations. We're implementing new legislation where we're going to define a framework and guiding principles for how we seek to answer these questions. In a big corporation, influence and advocacy are everything. Bottom-up efforts very often lead to burnout. You can have passionate people trying to advocate for change, but if you don't have an advocate, you will not succeed. Plain and simple.

You need influence! To get influence, you go to the top, build your influence there, and get a top-down mandate to do things differently.

Stakeholder engagement isn't something folks just do. There's a competency required, especially nowadays with our environmental justice communities. If you show up as a utility to a public forum without understanding what environmental justice means, what distributive, procedural, and reparative justice mean, then you won't get anywhere.

If you show up as a utility that hasn't built trust with your communities and you just start presenting solutions to people, you're not going to succeed. People need to weigh in on solutions to their own problems or you will meet pushback.

You need to understand each community's relationship to these issues. There's a historical disadvantage that makes it difficult to even engage them in decision-making.

What works is to partner with environmental justice community advocates and pay them for their time. We hire them to recruit, develop curriculum, and analyze insights. We compensate folks for their expertise the way we would with any other technical resource. Community organizations don't trust us. There's a history of tokenization. You say you're partnering with someone, put it on your website, but haven't achieved anything together; then it's just a headline.

There's the reality that your organization's composition doesn't reflect the communities you serve. If you don't have employees with lived experience, how could you possibly design for these communities?

If you have an all-white male leadership team at your utility, that doesn't read as though you understand the nuances of engaging with communities.

There's an internal piece and an external piece. You have to have recruitment and development internally and advance those who can represent these perspectives. That needs to be done authentically and in lockstep with engaging our communities externally.

There's so much history here that I have blind spots I can't fully understand. All I can do is try to approach the situation humbly and with curiosity. If we want to truly innovate, we need to show up with our communities to the commission jointly. The idea that you're going to put out a proposal, invite stakeholder feedback, and have it go according to plan is folly. You'll drag it out for 3 years and it'll be watered down.

Do the hard work. Engage your partners, your communities, and present it jointly to the public utility commission. The commission is going to be pleased. You'll have an easier time if you've done the heavy lifting.

I feel good about that message. It's one of partnership. No individual can create an ecosystem alone.

I'm optimistic. The only way we're going to get to new solutions to meet climate urgency is by partnering. If we continue to fight over methodology or cost-effectiveness, we're never going to get it done.

Do I think we're moving fast enough? Absolutely not. Do I have an appreciation for why we aren't getting it done fast enough? Yes. We don't turn on a dime.

I feel as though I can make a greater impact from within a utility than if I were outside. I've tried the public sector, startups, and consulting firms. I would rather put my effort into affecting change internally and advocating for new approaches.

That's what I see as my role. As my opportunity.

The Energy Access Advocate

Southeast

I'm in the energy efficiency space with a policy bent as opposed to a technical, analytical bent. I got hijacked into this career by the amazing people that I met in graduate school. Even though I am not an engineer, there are so many aspects to this work that could use folks with all kinds of skills, and I found my niche.

It was the very beginning of decoupling and the changing profile of utilities. The piece you could sell was that you don't have to give up on the service: your lights, the hair dryer, your dishwasher. You just have to get more advanced versions and consume less energy. Now, I have narrowed the scope of what I'm working on. I work specifically in the cooling sector: trying to solve the cooling conundrum. As the planet heats up, we need more access to cooling for many more people who are vulnerable to extreme heat. Unfortunately, cooling warms the planet because of the energy it consumes and the refrigerants it uses.

I'm on the efficiency side of air conditioners, refrigeration, fans, commercial refrigeration for vaccine and food. We need to make all of those as efficient as possible and as climate friendly as possible. The projections for how many more fans or air conditioners there are going to be in the marketplace in the next 10 years are terrifying. I'm in philanthropy so we make grants to try to achieve our targets. We are very committed to raising the floor through minimum energy performance standards by removing the least efficient items from the marketplace.

The standards are really important as a mechanism. If you make them more stringent, the bottom drops out. That's just what happens when you put a standard in the marketplace.

Right now we have a strategy with four priority countries, but our scope is global. In the first phase, we worked in 50 countries. In our second phase, we are focused

A. Kelly, *Good Work*, Women in Engineering and Science,
https://doi.org/10.1007/978-3-032-27547-9_21

on countries that either have the highest emissions from cooling sectors currently, like the United States and China, or countries like India and Southeast Asia where the demand for cooling products is growing rapidly.

Philanthropy is a catalyst role. You think strategically about where the opportunities are and where industry or government is not performing. You try to be the catalyst for bringing the good things forward. You're really hoping that through your catalytic grant that the government or private sector will take over from there. Cooling has only been a sector that has had attention for a short period of time. Cooling has never been its own thing. It started about 5 years ago when the predecessor of my program got a $50 million grant from 17 funders and made cooling a thing. We face resistance with this. The typical one is from industry. They say they can't do it; it's too expensive; it would take too much time to switch out production lines. They push back pretty aggressively.

In small countries like Thailand and Vietnam, they have small-scale manufacturers that provide local jobs. If they're not capable of competing at higher efficiency levels, they get worried about domestic manufacturers having to close down. Job loss becomes a political hot potato.

It is typically a negotiation of how far you can go toward the technical potential. The level of ambition is watered down because of these perceived barriers. The industry always has a knee-jerk reaction. They say they can't do it, no matter what it is.

For those of us trying to get the most CO_2 tons saved, it just adds time. Eventually, we'll get to a more ambitious standard, but we don't have time in the climate dilemma. There are days I'm terribly disappointed. Here's an example. I was asked to prepare a graphic that explained what a minimum energy performance standard was. This is the baseline introduction I've been doing for 30 years. You would hope that by now people have an understanding of energy efficiency concepts. It's painful to have to go back and explain it all over again.

I also know that we'd be worse off if we hadn't done all the work over the years. Great strides have been made. I tell myself that because it's hard to think of an entire career that didn't have the results you had hoped for.

There were various phases where I really thought we were going to succeed at arresting climate change. I was in Rio in 1992 at a summit, and it felt like we knew how to increase efficiency and get renewables online and that we were going to stave off this big existential planetary crisis. And I've been there several times in my career, and it just keeps getting worse. So we did something wrong. There's something we did not do.

It's really hard. It's really hard. Some of this is my personality. Other people remain very optimistic and hopeful. They see new progress every day, and I'm not saying that I don't. It just cumulatively doesn't seem like enough. All renewables that we have brought online globally today have been outpaced solely by the demand for cooling.

When we talk about diversity and inclusion, I feel strongly. We need new voices at the table. Not only culturally diverse, but age. The youth need to be at the table because this is their future. They're very invested and they're looking at it differently.

Those of us who are seniors, I'm putting myself in that category, we failed them. So why do we think we still hold the reins on this? It's time to really shift.

I'm not saying I'm not passionate. I sure am. I just recognize that it didn't get me nearly as far as I had hoped. I thought we were going to meet the 1.5 degree C. Now we are wildly off track.

What we've learned is that things are accelerating faster than we thought. We are hitting some of those tipping points faster than the models told us we would.

I think the approach needs to shift. We can't fly to a big conference every year. We did that for 20 years; now we need to find something different. In the sphere of philanthropy, something amazing has happened. Climate used to be the teeniest little slice and it has grown a lot. A grant of a million dollars doesn't make a difference, but it can be catalytic. To do something bigger than it is.

I'm willing to say I didn't do enough. Let us talk to the people who think we can leapfrog and really accelerate everything we need to do better.

I have had an amazing career. It's filled with passionate people who are so deeply invested. They believe that their work is important, even if it's just baby steps and not the whole enchilada. And they come back day after day and do it again.

It's been a really incredible community to be a part of. I love the people I work with.

The Watershed Manager

Midwest

I work for a state environmental agency that seeks to reduce pollution through watershed management. We are in a heavily agricultural state, and we are trying to adapt people's agricultural practices to be more sustainable to reduce nutrient and sediment runoff.

The kind of pollution we are trying to reduce is non-point source pollution. It's insidious: things like stormwater runoff or agricultural runoff coming from multiple different sources rather than the end of a pipe. On a landscape scale, these issues get concentrated when they flow downstream. Everybody lives downstream, and all the people and organisms that live in a water course are impacted by what's going on in their watershed. We try to incentivize folks via Clean Water Act funding to adapt their operations. I work with a program that sponsors organizations in watersheds. The state and federal governments have prioritized prolonging the lifespan of federal reservoirs built in the 1950s-1970s. Over time, reservoirs fill with silt. That impacts recreation and drinking water. Recreation has an economic impact, and those dollars go to wildlife and fisheries funding. Drinking water is a big issue. Communities downstream depend on those reservoirs.

The sponsoring organizations hire a watershed coordinator full-time that works upstream of the reservoir to try to change agriculture on the ground. One practice is cover crops. There are traditional farming practices like "till it black" that encourage tilling 100% of the soil and adding herbicides and pesticides. That culture of how to treat soil is passed down generation to generation. Farming is getting more expensive. Farmers have giant loans from paying for equipment or land, and everything they use is exceedingly expensive.

A. Kelly, *Good Work*, Women in Engineering and Science,
https://doi.org/10.1007/978-3-032-27547-9_22

We are trying to do regenerative agriculture. It's like sustainable agriculture, but that's a buzzword that scares a lot of rural folks. Regenerative agriculture uses cover crops to add nutrients to the soil. We also use cattle to add nutrients and reduce runoff, which will increase the viability of farming operations. I am not a watershed coordinator; I am a project officer. I dole out the funds from the Clean Water Act Section 319, which is devoted directly to addressing nonpoint source pollution. We administer the grants. We look at individual projects and ask if it's worth funding using our limited conservation funds.

Grasping the scale is difficult. There's also so much turnover. We've had half my section leave. There are 8 of us that oversee 15 large watersheds. As people have left, I've gotten more responsibility. I'm now responsible for 5 of them. But I'm losing count because so many people keep leaving. There's not enough money to go around. We are trying to build relationships with agricultural producers, and it takes a long time. We are trying to change long-held beliefs about how you do business. When we talk to the community, there's a misconception about our goals. People forget we're not actually trying to help farmers, but rather to protect water quality.

We can't pay for one farmer to buy all new equipment because we would use up all our funding on one project. In one situation, we had an operation right next to a creek. We wanted to move it up the hill. But the original operation was already out of compliance with state regulations. We can't pay to bring the rest of the operation into compliance or expand capacity. One main type of project we pay for is virtual fences. Think invisible fences for dogs, but for cattle. We also pay for alternative water sources. Imagine land that has access to a river. Cattle drinking water out of a stream are also defecating and urinating into the stream, causing pollutant load downstream. We establish fencing and an alternative water source to move that hoof traffic elsewhere on the landscape where it can be filtered by grasslands.

With virtual fencing, the farmer can draw out on their phone where cattle can go. They get used to the beeps just like dogs, and learn that a warning beep means a shock will follow. There's also the hope we'll use this technology to distribute grazing evenly and prevent problems from overgrazing. It's tough to get this work done because there's so much turnover and raises are hard to get. We end up getting folks right out of school, but as soon as they realize there are no raises or cost of living increases, they look for other work. The coordinators I oversee at nonprofits make more money than I do just because they don't work in the state agency.

It makes me philosophical about the impact I'm making. I got into this field for a noble reason. You can see all the horrible things going on and you know more about the issues than other people, and it can get depressing. I hope folks coming into climate change and conservation can get past this.

Part IV
Starting and Building a Climate Business

People don't know their heroes. I didn't, anyway.

These seven chapters are about people who bet on themselves. They started consultancies, founded companies, and built something from nothing. They figured out how to translate between jargon and meaning. They did the work that nobody else seemed willing to do.

One of them started a blog to answer questions she found interesting. It snowballed into a consulting practice.

One of them was 9 months pregnant when she got promoted to director, then came back from leave with her priorities completely shifted. She started a company with her business partner and a bonus from her last job.

That same founder built a nonprofit to train the next generation's green workforce. She remembers being in front of high schoolers and wondering "Do they think I'm cool?!" Of course not. It doesn't matter.

We'll hear from the only female engineer at a steel mill who now runs a company with her own patented, compression-based system cutting methane and greenhouse gas emissions from natural gas.

One of them grew up in Iran, where oil money fueled corruption and war. Now she runs an evaluation company and questions everything.

And we have a classic 1970s free spirit, starting her career in a Volkswagen bus with her glassblower boyfriend. She eventually figured out the money for a PhD, and ended up leading the energy evaluation industry, mentoring the minds leading change today.

I'm not worried about teaching people how to care. People care. I'm more interested in letting people know that they're allowed to care. That there's a way in.

The Program Designer

Mountain West

I moved to the Mountain West in the early days after graduating from undergrad. I had moved to the Virgin Islands for a couple of years and made a lot of money. I was just going to drive around the country and decide where I wanted to live. I knew I didn't want to live in Michigan, where I was from. I had a friend out West, so I picked her up and we drove all over the place. At the end of that long road trip, I stayed. I was tired of driving. When I first came here, I was a bartender and a waitress. Then I went to grad school and got a degree in environmental policy and management. While I was in grad school, I got a job with an energy services company. That's what got me interested in energy efficiency and sustainability. From there, I dabbled in different jobs: renewables, climate policy. I've covered all the different aspects of energy and climate.

One of my early jobs was in a consulting company focused on climate finance under the Kyoto protocol back in the early 1990s. I really liked consulting: the pace of it and the diversity of different types of work. Also, your ability to eat what you kill. I shouldn't say I always get to eat what I kill. Sometimes I get told what to eat. But I always want to go after program design RFPs because they're more interesting to me.

Those projects are fewer and farther between, and they're smaller and shorter. A planning project typically takes 9–18 months from start to finish. But if it's a five-year plan, you go away and come back 4 years later. You win a proposal and then you get to do the work. I liked that aspect of consulting within the clean energy space. My company covers a lot of different things, but it's all this social good, climate-driven type of consulting.

A. Kelly, *Good Work*, Women in Engineering and Science,
https://doi.org/10.1007/978-3-032-27547-9_23

I am a practice lead within the energy sector. My practice is called program design and performance. I lead a staff of about a dozen people. We work with utilities to help them design energy efficiency programs on the front end and evaluate them on the backend. On the design side, we're largely working within the parameters of energy efficiency resource standards in whatever state our clients are in. We're helping utilities write an energy efficiency plan that gets submitted to regulators.

The plan outlines how to achieve statutory goals established by the legislature. That involves designing programs, figuring out who programs will be offered to: residential, low-income, commercial, small business, large business. What kind of offering: incentive, upstream, downstream, midstream, energy audits, technical assistance, appliance recycling. There are a lot of different flavors. You bundle those into a portfolio that makes sense. We're also doing the math: forecasting what each program will achieve for energy savings. That's a modeling process where we often do different scenarios looking at what happens if you set your incentive level at one level versus another. It all drives participation, and participation drives savings.

There are two main levers to influence participation in an energy efficiency program: incentives and marketing. You offer more money, and more people are going to participate. You take a trendline and run it through scenario modeling. Each state has its own rules. You usually make a plan for 3 to 5 years.

After these programs have been running for a year or two, we go in and figure out how they're performing. There are two pieces: the impact side (counting kilowatt-hours and kilowatts, how much was saved, did it do what you thought) and the process side. A process evaluation looks at: could it be better? How did customers perceive the program? Did they have a good experience? Was it efficient? Did you communicate well with your trade allies delivering the programs?

I'm a principal here, so on most projects the buck stops with me. Most of the work to get to the end game is done by other people. We run a lot of surveys. Other people design surveys and program them and field them and analyze the data. Most of what I do on the evaluation side is reviewing, interfacing with clients, making decisions, pointing people in the right direction. On the program design side, I tend to get more into the weeds. That's important because it's getting delivered to regulators and utilities want to get it right.

I write testimony that gets submitted to the commission with energy efficiency resource plans. I'll review and respond to rebuttal testimony from interveners and sometimes serve as an expert witness if those plan cases go to hearing. And I sit on a lot of phone calls.

Right now I have three big projects. I have two planning projects at different stages. We're just getting started on one: starting to put the pieces together and see what the portfolio structure's going to look like. We'll go through a stakeholder engagement process, meeting with stakeholders who will eventually be interveners in those plan filings to get their insights early. We try to work with them ahead of time so it doesn't turn contentious. It's just more engaging to be upfront and work together. Sometimes they're successful and sometimes they're not. It's like a court case: the two parties try to come up with some compromise. If they can't, it goes to a hearing in front of commissioners who decide.

On the other planning project, we've already submitted the plan and we're in the interrogatory stage. You submit your plan, and any intervener or commission staff can send you data requests. You have to respond in writing. It's very fast (5 days to respond), and sometimes we get 20 requests. That does not get delegated to anybody.

My third client is probably my biggest. That's an evaluation client where we evaluate their residential portfolio. They have 14 programs. I have a team of probably 40 people working on that evaluation. I sit on phone calls and try to move the ball forward and review reports and deal with problems that come up, which are pretty much nonstop. I shouldn't say problems, but requests. That's one of the things I like about consulting: every day is different.

The part of my work that I enjoy the most is the planning. It's like putting a big puzzle together. You're shooting at a target. We have to save this many kilowatt-hours. In most states, there are other little requirements in the regulation you have to hit. These thousands of measures have to fit into programs that make sense that customers will want to participate in. The programs need to achieve all these different targets. That's the part I really like.

Just today, I was working with a client who said, "We want to shake things up, we're tired of the same old." They'd been running programs for over 30 years with 18 very standard programs. They were tired of it, so I restructured their whole portfolio to eliminate sector designations. Instead of residential, non-residential programs, they're just big programs that include all sectors, based on the delivery function. It's completely different than any other utility has. One reason I wanted to do it this way is because it's become increasingly hard to achieve cost-effective residential programs because lighting's going away and equipment standards have improved and gas prices are low. Combining the sectors pushes out benefits and costs among all different sectors and makes the whole thing more cost-effective.

I said, that's the fun part of my job. It's fun to come up with new ideas and be creative and solve the puzzle. We've got these problems. How are we going to figure out how to overcome them?

The impacts are sometimes hard to see. That's why I'm in this job in the first place: to create an impact, help make a more sustainable world, and hopefully have trees for our grandchildren to enjoy someday.

It can be discouraging. But in the last few years, it's started to feel like things are moving in a better direction. There were many years where my paycheck was being paid by a coal-fired utility. That didn't feel like I was having an impact. But you keep remembering that you're working with utilities. Sometimes I tell myself it's better to be working with "the enemy," the coal-fired power plants, because you're making an impact where it's most needed. It's better than trying to convince tree-huggers to hug more trees.

It feels better when I think about it. Getting utilities to be more energy efficient and doing more for their customers and helping low-income people improve their homes that are more comfortable and cheaper to live in. That's all good stuff. Some days, I'm just writing reports that are going to sit on a shelf. But taken in total, I do feel like there's an impact. I'm doing my little piece.

I've been in this since the early 1990s when the majority of people didn't believe climate change was real. That has shifted. Finally, people can see it from their own front door. The corporate world is starting to catch on. More people are requiring that businesses they work with be more sustainable. Some of it's greenwashing, but a lot of it's real. Public opinion has started to turn. The resource mixes are getting better. Most utilities have gradually increased renewables in their fuel supply mix. A lot of utilities have made commitments to be net zero within the next decade to a decade and a half. Sometimes it feels very incremental, but in total? Progress is happening.

The Methane Capture Entrepreneur

Mountain West

Our overarching goal is to support the natural gas ecosystem in reducing greenhouse gas emissions. We are doing this safely and reliably through what we believe to be an innovative commercial solution, a patented compression-based system that minimizes methane and greenhouse gas releases. It all evolved out of customer discovery. As you start to develop a product, launch it, and get into the marketplace, your customers tell you their pain points.

We are first and foremost focused on pipeline operations and maintenance. This is everything natural gas utilities are doing to shore up their infrastructure and make it as clean as possible. I believe the natural gas grid can be just as decarbonized as the electric grid if we apply all the right efforts and investments. Secondly, we are mitigating emissions through natural gas as a transportation fuel. Our first product to market addressed that infrastructure barrier around refueling with natural gas, and now renewable natural gas and hydrogen. Third, there are industrial operations, capturing emissions that would otherwise be released into the atmosphere. Capturing and reusing. Our customers receive expanded environmental benefits and reporting from our work. It's not greenwashing. It's real.

We just finished a deal with a major equity sponsor. Now we can offer purchasing, leasing, or services for our products. When we inked this deal, we were living our pandemic lives. The pandemic was not kind to us. We are a hardware company. We are a device you put in the field to make stuff happen. We are not software. We are not digital. The device is a big piece of equipment. You tow it behind a pickup truck. There are 800 parts. A lot of them are proprietary, machined, requiring sophistication: piping and valves that go into a compression-based system. It's hard to sell

A. Kelly, *Good Work*, Women in Engineering and Science,
https://doi.org/10.1007/978-3-032-27547-9_24

something like that over a Zoom call. It's hard when you can't get customers out to see it.

For the last 2 years, my job was fundraising. We were blessed to get non-dilutive government funding in the beginning from ARPA-E and the Department of Energy. We raised a little private capital, about three and a half million, mostly from angels. It took years. Until we inked this deal, my day in and day out was fundraising.

We just finished our hundred-day plan, a methodology our new equity sponsor brought us. They said when companies receive a massive injection of capital, especially coming from a bootstrapping background, they need to adjust to the new mindset of "go fast." For years we've been making personal sacrifices, and all of a sudden the back pain is gone, and there are millions of dollars available. Go fast.

For the last hundred days, I've been checking in with every part of the company to figure out what they need. Writing job descriptions. Putting budgets together. Fielding inquiries and customer excitement. Calibrating expectations because we don't have a product to turn over yet. That's a good problem, customers calling and wanting your product. But we are dealing with supply chain problems. We used to make products in 3 months. That's not our reality anymore. We're adding 20 people to our team of 15.

I'm the CEO and co-founder. Just this week we had an all-company meeting, and it's my job to set the framework for our culture as we try to go fast. I'm involved in hiring every person. I don't get involved at every stage of every interview, but I interview everyone in the end. It's important we have the right fit. We have representation from all aspects of the political spectrum, and I think that's great as long as we are all focused on climbing. I'm a climber, so I make climbing analogies. We are climbing the same mountain. I don't care who they voted for or if they didn't vote.

I wish the rest of America could see us. You can work together when you establish shared values and beliefs about what's happening in the world. We don't have to talk directly about climate change and reducing GHGs. If you're going to be cutting into a pipeline to perform an operation and there's gas in that pipeline, you have to evacuate the gas. In common practice, you turn the valve and let the gas go into the atmosphere. It's not hard for anybody to see that shooting gas into the atmosphere is wasteful. There's nobody who thinks that's a good idea or good practice.

And there are crazy pictures now showing massive methane leaks with infrared cameras from satellites. There's nobody who thinks that's okay. I don't believe in either side. I think there are wackadoodles on both sides. I try to find the vision in simple terms that would allow everybody to come to the table. We should all be climbing this mountain together.

Public education is a big part of my work. Everything I present depends on the audience. If I'm giving a very technical presentation to pipeline operators, I acknowledge that they need tools that work for them. They don't want to do this, but they're living with time constraints and money constraints and pressures. They need tools, and they've got bias depending on how long they've been doing it. I acknowledge what they do and show respect. Their work is dirty work, and they need solutions that work for them.

There's a general audience on the West Coast where people are banning natural gas. Schools are making decisions to electrify everything, even though their utility offered to supply renewable natural gas. They're pushing to electrify everything. But what provides their electricity? Where's their power coming from? And when the grid goes down (which it does quite often), guess what? The natural gas grid doesn't go down. In Texas during winter storm Uri, a lot of people survived the deep freeze because they were able to heat and cook with natural gas when electricity went down. It was horrifying. We live in a first world country, and there are decisions made to electrify everything, but where does the electricity come from?

You know how blessed we are that we have two grids independent from one another? You can decarbonize and talk about renewables for the electric grid: geothermal, hydro, solar, wind. On the natural gas grid, you can also introduce renewables. You can blend renewable natural gas one for one, and you can blend in hydrogen up to a certain percent. There's a lot of work happening on that front, and the Europeans are more advanced. If you introduce renewable natural gas and hydrogen, you can get that entire energy source pretty close to net zero.

You have to stay focused on how to get to net zero. It's going to take a whole combination of things. You're not going to get there by cutting off entire sections of energy, because renewables aren't going to satisfy 100%. We know that today, but people don't quite see that math.

Take dairy, for example. Using dairy biogas or any kind of methane biomethane source, we can actually become carbon negative. It has a carbon intensity score that is carbon negative. People don't understand that. There are charts showing that when you burn methane coming from a renewable source, it can be carbon negative, unlike an EV being charged off of coal.

Things are changing. I feel like with ESG, the whole world is just catching up. We've been doing this for 8 years. There are three aspects of ESG. The environmental is where you implement the technologies and procedures to address your objectives. The social piece is how companies treat their people, communities, and implement DEI. Corporate governance needs to be not just greenwashing. Use reporting, third-party verification, and audits. Show who is really writing the checks.

I am trying to recruit a more diversified workforce. Right now, we have 15 people and 5 women. A third of us are women. We've got two women in the executive circle (me and my CFO) and we're a three-person executive team.

There weren't a lot of women to look up to when I entered the energy space. In my generation, women are few and far between. Even today, when I go to pipeline technology forums, probably only 10% are women. I feel like women need to step up and help each other. We had a little group of women in my area, and I told them my commitment was that their emails would go to the top of my list. Men are really good at reaching out to their networks, to their dad's networks. We know only 2% of capital goes to women. Women need to learn how to help other women.

I worked in a steel mill. I was the only female engineer. The person who hired me thought I was gay because I was an engineer with short hair and wore pants to the interview. I never experienced cattiness because I was in my own field. I got so tired of being the only woman in the boardroom. The scary thing is I don't see women applying now. We have all these jobs open and we aren't getting women applicants.

Our company spun out of a university under an amazing engineering professor and inventor who was the original innovative genesis of the project. They had the original ARPA-E award. If the technology was going to be commercialized, we needed to spin it out. It goes to show your network is so important.

There are all these paths to 2050 to net zero. My hope is that we can embrace logic and reasoning and not throw out any solutions. All I know is what we're doing at our company is a piece of this massive puzzle, going after the low-hanging fruit of stopping methane from being released. Every time our equipment is used, I can go to bed knowing I have prevented a hundred thousand cubic feet of methane from being released. I can't solve it all. I can only play a little. I have one small piece of this big puzzle, and I only hope common sense will prevail.

The Climate Investment Educator

West Coast

My blog is about sustainable investing. It's called something catchy, a platform I started a couple of years ago to share information in an understandable way about sustainable investing or ESG investing. It's a growing sector, and people are more interested in learning where their money is going and how they can try to do good with their money. They want to align the money they're putting in a bank or in a 401k with their values. But if you look online, the information is very focused on institutional investors or investors who have been doing this for a long time. I wanted to help people shift money away from the bank because banks are huge funders of fossil fuel projects. When people go on my blog, they can get information on funds: their fossil fuel investments, how they do on gender equality and racial equality. I'm trying to make that information accessible to everyone.

At the start, I was thinking about doing it as a business. There's an amazing book called Leap Frog about how to become an entrepreneur, focused on women and non-binary entrepreneurs. One of the ideas was to simply punch something you don't like. Because it didn't end up being a business, I just write in the evenings and on weekends. I happened to get a day job that I really loved, so it was fine that my blog ended up being a side project.

I worked through the types of investments most people have. I was frustrated because I was trying to find information about sustainable investing, something I've worked on for decades. I started doing sustainable investing when I was in my 20s, and it always took me so much time to find the information. I knew I shouldn't complain; I should just do something to make it take less time.

I started with bank accounts because that's something most Americans have. I track when new sustainable banks open, review them, and write blog posts. I also

A. Kelly, *Good Work*, Women in Engineering and Science,
https://doi.org/10.1007/978-3-032-27547-9_25

look at 401k plans. I try to let people know they do have a say in what goes into those funds because it typically seems like companies give people options and that's all they can access.

Then I move into the stock market. I've had different people write on my blog about everything from tiny homes to putting solar energy on your house. It's a passion project. I'm very passionate about sustainable investing. I'm climate focused and want everywhere my money goes to be somewhere I feel good about. If I'm buying clothes, I want them to be sustainable fashion or from a thrift store.

Some people write blogs off the cuff, but I really enjoy the research and find it important to write something educational. You need to understand the subject and background.

Let's take banks. We have banks that advertise themselves as clean banks. They might just have a clean bank as the front layer, but a more traditional institutional bank behind them. You have to dig into the fine print to understand who is behind them and how they've set up their business model. I want to enable the reader to learn about the different layers because they are complex.

Let's say we want to push on ESG aspects of a fund. If I'm paying extra fees to invest in an ESG fund, I want to know my fund manager is taking real action to push on the environmental, social, and governance aspects of all companies in my portfolio. I wrote a post discussing the benefits of racial and gender diversity at a company and wanted to understand what we, as investors and part owners, can do to push companies to take action. I put in a draft letter people could send. I personally sent it to a bunch of companies, and I got some really great responses. One fund manager wrote a detailed response about what they were doing to push companies to improve diversity, what they were asking, and what they were tracking. That's exactly what I want to hear from a fund manager if I'm paying that extra fee.

We have such a long way to go. There are so many different pieces we need to create for a climate framework. Europe has done a lot more than us. Although we shouldn't have to push companies to do this, the transparency and standardization will eventually be good for business. Smart companies have already been paying attention to climate change and emissions. It's also beneficial for investors. I try to remind people that when you invest in a company, you are a part owner. They have a fiduciary responsibility to operate in a way that gives the best return, including financial and climate risk.

I am just one voice of many. I have probably impacted a smallish group of people, but that's where it starts. If I impact a small group, they'll talk to people. I've had so many friends and colleagues reach out with questions because of a single blog post. It makes me feel like it's doing what I hoped.

The blog is just my side project. My day job is at a company aiming to enable the clean energy transition. We are enabling buildings to be virtual batteries. Buildings can shift and reduce load to help out the grid. Normally, you have an electric meter between the grid and the building, and grid things happen on one side and building things on the other. We are trying to cross over and allow buildings to operate in a way that lets the grid get cleaner while buildings operate more efficiently. I'm the director of strategy, which means I do a lot of different things. I'm a gap filler.

I do a lot of writing at work: strategic documents like business plans, proposals that require thinking about where we're going as a company. I help with sales, do some marketing, and work with our R&D team doing cool experiments around demand flexibility. A big part of my work is storytelling. The work we do is complex, so we need to think about how to simplify that story for different audiences while also being compelling to the utilities we work with who understand the details.

We are taking what utilities know about energy efficiency and demand response and merging them to see if we can do smart, data-driven energy efficiency that enables us to learn about buildings.

We're trying to figure out how to do wide-scale demand response and increase flexibility to prevent outages. There was a power outage in the Pacific Northwest last year because of a heat dome. People lost power for 4 days. We can work with buildings, solar, storage, and EVs and look at all that data in our software. We have a dashboard showing what's going on from the building perspective, solar perspective, and utility perspective. We pull in data from different sources and figure out how the pieces work together.

We work with national labs to run experiments on buildings to understand what happens when you adjust systems, how that impacts both the grid and building occupants. It's really important to us that we don't negatively impact the people in buildings. We've also staffed a 24/7 call center to respond to issues around the clock.

My day-to-day is pretty normal. I make a list at the start of each day with three things I'm going to get done. Other things can get done on the side. Right now I'm in a lot of meetings because I'm involved in many different things. I try to block at least 1 day a week with no meetings so I can write and review. We have an amazing writing group at the company. I'll write a first draft, then ask everybody for input. We have such great brains! I'll get amazing input and make a much better second draft. Somebody has to do it. I'm a former consultant, and consulting is pulling information from different sources to give people a starting point, then taking critique and making it better.

I wouldn't be here if I didn't think it would have significant impact potential. The energy transition won't happen without all these different solutions. We need storage, controllable EV charging, different types of generation, and buildings have a major role. Buildings generate 40% of the world's emissions. The potential for impact from making buildings run more efficiently, not to mention more pleasant to be in, is huge.

We're already seeing that we aren't doing enough fast enough. As someone who has worked in developing countries, I've seen the impact of changing weather patterns and extreme heat. My heart goes out to all the people constantly impacted by extreme weather. Tragically, we are seeing our failures. I work on climate change from a business perspective because there are so many economic drivers to reduce emissions and prevent climate change from being truly catastrophic.

You have to be optimistic to work on climate change. Sure, you also need to be realistic. We are seeing changes from solar and EVs. The growth over the last decade is astounding, and it's going to continue. There are pros and cons of every technology, but this growth is important, a sign of hope.

Taking action on climate change has so many side benefits. It helps health, reduces poverty, helps gender equality. It helps everyone to transition to cleaner energy and to be more efficient.

The coolest thing about climate change is that there is space for absolutely everybody. Everything impacts climate change. Whether you're in the restaurant industry thinking about sustainable food production, in marketing wanting to work for a clean technology startup, in a factory figuring out more efficient ways to run, there's space for everybody. Everybody has a role to play in the climate fight.

The Energy Consultant

Mid-Atlantic

I have my own energy consulting firm. No two of my projects or clients are the same. Sometimes people need white papers or blog posts, and that's simple enough. I have the skills to do the research, interview their teams, and write. Then I have other projects that are more ambiguous, and I have to figure it out as I go. Sometimes I do energy models or lifecycle analyses for companies who don't have expertise in tracking their carbon or looking at the impact of their energy use. They bring me in. I'm the person who knows what questions to ask and how to figure out what questions they need to be answering. It boils down to being a jack of all trades and a subject matter expert in the world of energy sustainability. I come at it with an engineering background from school and a lot of exposure to energy policy. I take my technical background and make it easy to understand.

The first decade of my career was spent bouncing through a couple of major consulting firms in Washington, DC. That's where I first got exposed to the Department of Energy and their work. I worked with a small company, then with a big one. It was good exposure, good training. At the end of that timeframe, I felt myself getting restless because I got into energy and sustainability to make an impact. At larger consulting firms, I felt like a cog in a wheel and couldn't see the bigger picture of the work I was doing. I didn't see a way to put my name or stamp on any final product because I was just part of a bigger machine.

I started a blog in the energy space with my perspective. I thought I should start answering questions I find interesting that maybe other people would find interesting too. I wanted to use it as a way to connect with people. I was too low level to meet people across the industry. I thought if I put out a blog and shared it on social media, I'd be able to meet people and have conversations. It did well, didn't take off

A. Kelly, *Good Work*, Women in Engineering and Science,
https://doi.org/10.1007/978-3-032-27547-9_26

or go viral, but it scratched an itch and opened doors. Then I started to wonder if someone would pay me to answer questions for them.

It snowballed. I wasn't just writing papers anymore. I was doing analyses. I ended up spending all of my personal time, nights and weekends, working on it. It finally got to a point where I had long-term clients and sufficient interest from others that I was turning away work. My wife agreed, and I jumped all in. That's what led to where I am now. It's been about 3 years, and I'm back to having more work than time. That's a great place to be.

My day-to-day is what you'd expect. I'm typically in my home office, mapped out from 6 a.m. to 6 p.m. I don't want to do that forever, but I'm putting in the work now so hopefully later I don't have to. I spend the first few hours on regular daily work for long-term clients. I'm a community manager for an energy publication, looking at submitted content, answering emails, addressing projects. Then I'll have hours in the morning and afternoon for project work. There are always meetings: daily check-ins with some, weekly with others, ad hoc meetings with potential new clients or interviewing. The beauty of the work is it's variable from day to day.

One really interesting project was on electric scooters. Those scooters are in cities everywhere. Cities were starting to ask for sustainability analyses before issuing permits. I got involved in a couple of related projects. This was an area where I hadn't had institutional knowledge. Sustainability wasn't something they'd thought about because it was a brand-new business model. It was such a big challenge that I felt paralyzed. I didn't know where to start. In the end, a lot of decisions are dictated by the data available. If I imagine my dream dataset and flash forward a month to see what I have, it's very rarely aligned. These companies don't track this information because they didn't see the need. Data is expensive to track, or they don't know how.

The initial stages of building those models are interviewing people involved. I spoke with people purchasing scooters, managing repairs, making decisions about end-of-life. All these conversations led me to find what data I wanted. Sometimes I get surprised. Someone has the data and sends me the file. Other times they don't have it at all. I wanted to know how frequently certain parts were repaired. That data wasn't available. Instead, I asked for financial reports on repairs and worked backwards. There's a lot of desk work doing research and reading academic papers, industry papers, and government papers. Sometimes there are best practices. A lot of times I'll find one little part of a study useful, then another paper to fill in blanks. Once I have the information, I hearken back to my engineering knowledge on how to build a model. I do math and poke holes in everything.

I keep in touch with clients, letting them know what I'm finding and providing early-stage results. Sometimes they'll say what I'm making isn't what they want and ask me to pivot. Staying in constant communication is key.

Some lessons I've learned aren't ones I'd necessarily espouse to others. I've taken the approach of saying yes and then figuring out how to do the work. That comes from the hustle attitude. It's hard to say no to a project. I have limitations. If I'm offered work way outside my scope or not within my expertise, I'll say that. But if a client asks for a model I don't know how to make or wants to know if I have experience in a specific area, I'll lean toward saying yes. I'm confident in my ability to go out and figure it out. I have a network to ask for advice or to bring someone in. One of the great things about the sustainability and energy space is people don't look at their work as zero-sum. People are mission-driven. We all want to get to clean energy, clean transportation, a brighter future. When you reach out and say you're working on a project toward our goals, there's an openness. Strangers will call me, and I'll call them, and they're happy to sit down for half an hour. The lesson: fight your fear. This is a collaborative industry.

When figuring out how much to ask a client to pay, I almost always aim for time and materials and get paid by the hour rather than a fixed flat fee. I don't know if a job will be a lot more or less effort than expected. I also think it's advantageous to clients because they're not paying a fixed fee for something I can pull off quickly. There's trust to make sure time is being used well.

When I take a project, I need to think about how it's going to add value to my company, my brand, my profile. When I do content writing, I prefer if the client lets me put my name on the byline. Both my clients and I can benefit from that arrangement. Once I can credit myself, I'll put it on my website and link to theirs.

I had written an article on my blog about electric scooters that got attention. I was living in DC at the time, and I saw these companies filling up the city claiming they were a sustainable solution. Then I noticed they had big diesel vans picking up scooters at the end of each day and moving them around. I thought this might be greenwashing. I wanted to find out. I did some back-of-the-envelope calculations and put it on my blog. Based on information I found, I wrote that I didn't think it was fair to assume in either direction yet. I was one of the first people who ever tried to run the numbers on whether they were actually sustainable. Since then, there have been academic and government studies, but it's cool because my blog has been cited in them. That's what led to me meeting potential clients who wanted me to review what they'd be doing. Now I've helped scooter companies self-assess their sustainability and identify improvements.

They don't have a sustainability expert on their team. When they bring me in, I can look at their carbon footprint or help identify weaknesses in their environmental plan. I've turned away clients who wanted me to write about something that looked sustainable but wasn't. That's not how I want to contribute.

I'm not saving the world. I'm not doing anything on a grand scale. There are certainly people inventing new technologies or passing policies that make a huge aggregate difference. I don't look at myself as having that giant impact. When I think about impact, I remember a high school teacher talking about what it means to have a PhD. The goal was to come up with something that adds to the bank of knowledge in the world. It's going to be esoteric and narrow. He showed a graphic where a giant sphere of knowledge had a little pimple on it from your PhD. But when you zoomed in, it was huge. The closer you got to your area of work, the more valuable and impactful it feels.

I don't have a PhD. I don't feel like anything I'm doing couldn't be done by somebody else. I'm not breaking new ground. But I am slightly pushing things forward, through a blog post, working with a client and helping them share their message, or assessing their environmental impact so they can make better decisions. In the grand scheme of the climate change fight, it's a drop in the bucket.

That sounds pessimistic, but it's meant optimistically. I want my work to be that drop in the bucket. In the network I've built, I'm seeing everybody make those drops. That's how, if we're going to get where we need to go by 2050, it's going to be because so many people make their little impact and focus their life's work on solving this problem.

When you look past the climate doom feeling, all is well. I'm doing the part I can. I'm proud that the work I'm doing is moving us in a good direction. That's how I bring it.

The Capacity Builder

Pacific Northwest

When I was in college, I intended to study biology. After my first year, I escaped. I realized I liked people better than frogs, so I became a social psychologist and got a bachelor's in psychology.

How do you make a living doing that?

College was over, and I started looking for research jobs. The first job I had was at a public school doing program evaluation. I had never heard of evaluation, but it was all I could find. I worked part-time for a couple of years. Then I worked at a Construction Industry Research Board. I had an interest in buildings because my father was a contractor. I felt like I could look at the construction industry and fit in.

I did not like Los Angeles. I stayed for about 3 years. But this was 1978, the same time as the energy crisis. Because I had been working in construction, I felt the energy crisis really impacted me. I had read several books about reducing our consumption, and I had heard about global warming in a magazine called Co-Evolution Quarterly. At that point, we knew from the models that we were putting too much CO_2 in the atmosphere and that there would be consequences. I was pretty alarmed. We had natural gas shortages for a couple of years; this magazine was talking about climate change, and I was doing social scientist work.

Right at that time, the first building energy codes and appliance standards were being passed in California. Because I had worked in the construction industry and built their library on energy conservation and solar, I knew there would be big impacts. I went to a solar industry conference and learned as much as I could. I remember sitting there feeling out of place. "I'm a social psychologist sitting here taking notes so I can learn about how energy comes from the sun?"

A. Kelly, *Good Work*, Women in Engineering and Science,
https://doi.org/10.1007/978-3-032-27547-9_27

At this time, I was also looking into intentional communities. There was one outside of a small city in the Pacific Northwest. My boyfriend and I were members of a group looking into moving there. I moved and became a leader in the group. Eventually, I broke up with the guy and got out of the community. But I liked the area. I looked for a job and found that my hairy armpits coupled with sleeveless dresses didn't work for all employers, regardless of how environmentally forward everyone seemed. At this point, I was living in a Volkswagen bus with my glass-blower boyfriend, then husband. As an aside: when you're young is the time to really enjoy life. It wouldn't be easy now to live in a bus, but it was a lot of fun then. I'm glad I did it.

There was a recession which brought federal funding into play. I got a federally funded job to do a survey of downtown neighborhood residents of either single-room occupancy hotels or very ritzy high-rise apartments. That was interesting, and I felt like I was using my social science training. I still wanted to work in energy. Eventually, I got a job with an organization that did workshops on how to build solar greenhouses and solar hot water systems. I was finally in the energy industry.

Right around that time, the Northwest Power Planning Act was passed, a bill requiring four states to appoint people to a council that would oversee and develop least-cost planning for energy. Many people in the energy industry lost their jobs because federal funding disappeared. Ronald Reagan had been elected, and he famously said conservation was freezing in the dark. He took down the solar panels from the White House and canceled all the funding for regional solar organizations. All my colleagues saw the writing on the wall and went off to do other things, and I was left holding one remaining contract. We were supposed to install solar hot water systems on six low-income households. The households weren't interested anymore because of the "freezing in the dark" comments. The money was going away, and at the same time, the regional power authority was just starting its first energy efficiency programs.

I wrote a proposal to try to get some work with them, but I quickly realized I needed a PhD to get contracts like that. I didn't have a job, nothing. There was so much going on, but I couldn't get into it because I just had a bachelor's degree in social psychology and nondescript grassroots work. I didn't have any credentials whatsoever. I ended up going to the state school to get my PhD. I took the bus to school, and the state school was reasonably priced. I did have to borrow money, and there were days I could barely afford to ride the bus.

It worked. After school, I met a wonderful man who helped me make the right connections. After a couple of tries, I landed a job where we did evaluations conducted by a national lab. We worked on data collection for all of the residential weatherization programs in the Pacific Northwest. It was a small evaluation firm, and I consider that to be the beginning of my career. From my perspective, it was like I had managed to get into heaven. Never would I have thought I would end up doing evaluations for utility companies, but all the pieces fit together. I think the key thing was that I wanted to do research and I was willing to do it. The energy crisis plus my recognition of climate change kept driving me to what I wanted to do. My parents were not really happy that I kept sticking to my dreams. It took a while, but

I did get closer and closer. You have to have that persistence when it's almost too expensive to ride the bus.

The regional power authority, due to getting all of that funding right before Reagan, ended up being the launching pad for many people in the energy industry. Once these people were trained, New England started having power outages and brownouts, and there was a huge cadre of people who had been working in energy for 5 or more years providing brainpower to other parts of the country.

I worked alongside these folks for several years. After my PhD, I joined a California consulting firm. Then, in 2001, Enron happened. Afterward, the whole energy industry restructured. By that time, I had enough skills that even when my part of the industry imploded, I could start my own company and focus on what I was interested in. My old company wasn't interested in pursuing any more work, and I had already started working on proposals. The company let me take them, and I pursued them and won them. Suddenly, I had my own company. I knew I could do it. You can run a business in your house by yourself. Eventually, my company grew to 33 people. In 2019, we sold.

Doing this has been a lot of fun. I had to learn how to run a business. Here I was with my social psychology background working in engineering and helping utilities be more effective. There is so much to learn and understand. I learned about how people adopt energy technologies and how they interact with technologies and organizations. I took a class in running a small business. Although I eventually sold the company, it was never my intention to build a profitable business and then sell it. My intention was to have a company where I could enjoy working and train people to do the type of research I do. I wanted to increase capacity in my industry. I'm really happy that the majority of people who worked with me over the 23 years are still in the industry and still doing this work. They are making a difference. I went to a conference recently and there were ten of us there, more than I could ever afford to send. It's very gratifying to see that our intentions can be achieved if you have them, pay attention to them, nurture them, and don't give up.

This industry has been through phases: energy conservation in the 70s with Carter, then demand side management after Reagan. We have this notion that the market should take care of everything. In my view, utilities should do energy conservation because their customers love it. There has never been low satisfaction when you're getting energy services. Even though energy efficiency is a no-brainer for utilities, we still find boom and bust cycles with funding for programs. I have always had confidence that I would be able to morph with the changes. There are a lot of different strategies to reach our goals: grid modernization, electrification, energy efficiency, demand response. Those are only some of the things going on. Do we really need energy efficiency? It's not really cost-effective against nice solar plans. But on the other hand, you can't really have a good solar investment without also having energy efficiency.

The perception of climate change is the hardest part about this work. Once Reagan politicized the issue, it's been a horrible battle. There have been times when I couldn't feel comfortable talking to other people about what we do. They didn't believe it. In some ways, it's worse now. There's more hostility.

There are so many career opportunities. Some technologies, like carbon capture, seem like they could never happen. But even if that's true, it'd be a shame to give up because it doesn't make sense today. Maybe we will get the hydrogen economy. If there's any chance, then I think we ought to go with it. Let's be flexible to ride the wave but stick to the idea. The main problem is still the main problem.

The Workforce Developer

Pacific Northwest

I was a director of operations for a large energy efficiency implementation firm. I had worked my way up the ranks over 7 years. I handled internal operations and supported energy efficiency programs in 14 states and managed energy efficiency portfolios in 2 different states. I got to touch a lot of different energy efficiency program designs and to bring new products to market. I also did people management, leadership development, and HR stuff. HR was the hardest for me because I care really deeply about people.

It was hard to separate out when people would come to me with problems, not knowing if I could help them or not. In consulting, the work is contract to contract. Sometimes you have to make hard decisions about layoffs. I always had to know who I would cut if asked. I always had to have a list ready to go. I hated that part. I knew everybody's personal stories, but I also had to make the best business decisions. Layoffs usually come right before the holiday season, which is even tougher. Firings are easier because people know it's coming. But people don't know layoffs are coming, and I found it brutal.

Then I had a baby. I had set a goal when I was 26 that I was going to be a director by age 35. I was 9 months pregnant and got promoted 2 weeks before my 35th birthday to the director position. I got what I wanted right when my priorities were shifting 100% because I was about to have a newborn baby. I came back from my short-term leave, and my anxiety was ridiculous. Learning to balance the directorship and the baby was really difficult, and I knew I couldn't keep up my normal pace.

I started thinking about what I could change. But also, I didn't feel like I was seeing enough representation of women leaders in the energy efficiency space. I wanted to make sure women were represented. At that point I thought, "I'm ready.

A. Kelly, *Good Work*, Women in Engineering and Science,
https://doi.org/10.1007/978-3-032-27547-9_28

Let's do this." By this point my son was two and a half, and I ventured out and started my own company in 2017.

Most small business books tell you not to go into a small business with a partnership, but I did anyway. I chose a really good business partner, and we are yin and yang. I am in business development, outgoing, the face of the company. She is 100% operations, administration, paperwork, certifications, managing contracts, all the stuff I have a hard time doing. She also took on 50% of the risk. We have bootstrapped this business from the beginning. I had a bonus from my last job. I used the money and my network. I quit first, went full-time, and got the pipeline going. My partner was in the background getting certifications, the website, and the bank accounts. Then I went out to get contracts, and as soon as we had enough to pay for two people, my partner quit her job and came on full time.

When I started the business, I made sure that every day I woke up and took at least three steps to further my career. Whether that was finding new connections, even if I didn't feel like it, sending out ten emails, or making three phone calls. I just made small daily goals so I could get momentum and the ball rolling. It's amazing how that works. You just have to take initiative.

We knew we weren't going to be a software company. We were going to be a services company. Our business model was to subcontract with larger organizations and use our WBE minority spend in our industry. Utilities want to see diversity in their contracts. I hit up all our connections and got on supplier diversity lists. That was my first mode of operation. People knew my specialties: program design, evaluation, strategic energy management, data center programs. The folks that knew me already knew my reputation and that helped get a foot in the door. I also reached out to colleagues who had started their own companies and tried to find companies like software companies that could complement our services.

In the beginning, the ambition was to make money, survive outside corporate America, and promote women in energy. Our other mantra was that we could say no if we didn't want to work on a job. There have been times I've kicked myself for taking a contract that didn't align with my core values. My partner and I often talk about our core values. It's amazing what you start attracting when you focus on your values and narrow down your focus. Then you see an uptick in people interested in what you're bringing.

I come from a blue-collar family. We were expected to do our best with what we were given. When I rely on my confidence, intelligence, expertise, intuition, and creativity, I attract the right attention and the right contracts. It's actually a nice thing about aging, not focusing on appearances so much. The hardest thing is when inappropriate comments are made. I don't want to giggle at it because I don't know what to do, but afterward I kick myself because I didn't want to make them uncomfortable even though I was the one who was made uncomfortable. It's really hard for me to go to business conferences because that stuff still happens.

One of our first projects was working on utility allowances for affordable housing. In this kind of program, developers invest in energy efficiency or water efficiency and can increase their utility allowance. We made good money on that product. But I thought if we were going to make money on products like that, we needed a way to be a socially responsible organization that could give back to communities. We started a nonprofit off the success of that product in 2018, It's a 501(c)(3) dedicated to training the next generation's green workforce with equitable pathways to green jobs. We provide internships to high school students where we help them get their resumes and LinkedIns together, and we have an in-classroom component where students learn about green jobs. They learn about jobs in utilities and all the high-paying green jobs these kids need to know. We also teach foundational life skills like how to build a resume and turning your lived experience into resume experience.

We know in the professional world that birds of a feather flock together. We are trying to change the flock. We want a more diverse flock. We need to get people out of their comfort zones and interacting with different communities.

We launched our first in-classroom program and ran our first internship in the summer. We have impacted 165 students in just 1 year of in-person programming. Getting this to work is a lot about timing. I run our nonprofit like a for-profit, except the money goes back in. You need to pay people because they're working normal business hours. We partnered with a great sponsor and get funding through a foundation that produces a lot of curriculum across the country for teachers. They weren't seeing it get into the classrooms. We're the boots on the ground, so we took aspects of their curriculum and got it in there. We want to connect the people aspect. I'm not an educator, and there can be hurdles, but you ask questions and problem-solve until you get there.

I learn something new every single day. I talk to people, I listen to podcasts, I read a lot. If I need to learn something about policy or regulations, I'll ask people and have conversations. I had never worked in the classroom before I taught a 100% virtual class. I made so many mistakes. The biggest thing is asking for feedback when you know something is not right. Ask those around you for honest, clear feedback and try to take a deep breath, put your big old ego aside, and take that feedback and adjust and tweak until you get it right.

Getting back in front of high schoolers raised all my own high school anxieties. It's been a long time since I was in high school. When you're working with 17- and 18-year-olds who have been through the pandemic and everything is virtual, I still had the moment where I wondered, "Do they think I'm cool?!" Of course not! And it doesn't matter. My epic failure was trying to do too much within a given amount of time. I always try to cram things down people's throats. I had to learn to back off, revert to listening, and take it slower. I needed to ask more questions and give students things to research and come back with answers. I think the word is patience. Just sitting back and watching learning unfold.

The work I'm proudest of is this project, hands down. I'm not done yet. I'm just getting started. I look at it as an incubator for all my harebrained ideas that are not cost-effective to start. I work on them until they are in a place of cost-effectiveness so we can use the right budgets in the utility space. I feel like I have a design for a new program we could launch on the for-profit side that could be a nationwide program.

When I started at a utility in 2001, I was already noticing lots of people retiring, and we weren't getting institutional knowledge transfer. I feel like nobody cares about this industry because it's old and not glamorous. It's not Microsoft or Amazon or Google. But we still need people with new ideas. I'm glad the image is changing a little, but we still need young minds.

I have a mentor who is crazy positive. Like unicorns and rainbows positive. I called him once because I was in despair that no one was ever going to buy this stuff and the world was going to come crash down. He told me, "Back in the 80s and 90s, we found out that the ozone had this huge hole and scientists told us it was going to destroy the whole planet. You know what we did? We fixed it. The ozone layer has come back. We have reversed the impact. We can do great things if we just put our minds to it."

It's a great example that when push comes to shove, humans may wait and wait to solve a problem, but then they'll go all in when they really have to. That's how I feel right now about climate change. I don't know what the breaking point is. That's the scariest part because I feel like we're already there. But I know that humans are good to the core and that human innovation can get us there. There are so many smart people, and we can do amazing things if we put our brains together.

We haven't hit rock bottom yet. For some weird reason, as a society, we don't get it. But by making an impact and training the next-generation workforce, we can generate the spark, the one person who will make the crazy invention or service that gets us there. My part is to try to attract those minds to our industry that can put forth the revolution in the way we interact with our climate and our resources.

The Program Evaluator

Pacific Northwest

I come from one of the big oil-producing countries, and I grew up seeing how oil money fuels corruption, war, and violence. Oil gives governments too much power and creates dictatorships. It creates lazy citizens that expect the government to do everything because the government owns the oil. Other people think we should stop climate change for resiliency or security, but I do it to avoid wars. That's the cost of not fighting for clean energy.

My life has been impacted by climate change. I was raised in 8 years of war. The Islamic Revolution never would have happened if we used clean energy. None of that shit would have happened. I was born and raised in Iran. I moved to the United States in 2002 when I was 24. I was so surprised by the anti-nerd attitude here. We don't have that in Iran because it's a culture that puts a lot of weight on knowledge and education. I was surprised that there are anti-nerds all here.

Education is the most important and most powerful equalizer. Let's not even start talking about what it's like to be a woman in America. Growing up in Iran, of course I have opinions about oppression, but American women are oppressed too. They don't get close to science. They have a lack of confidence. We have very few technical women. I run a company. I worked for big corporations for 16 years before starting my own. I am trying to quantify the impact of all the different steps we take to reduce energy. I know we can do it more accurately than we are currently doing it. I have seen a lot of poor methodologies. Society can do this for a lot less money if we do it differently.

I want my company to catch the early changes, see the wave coming, and start paving the road in front of the cars. I've been seeing changes measured in

A. Kelly, *Good Work*, Women in Engineering and Science,
https://doi.org/10.1007/978-3-032-27547-9_29

cookie-cutter ways using the same methodologies from the 1970s. We need to take things to the next level.

I also started my own business for diversity reasons. The only way women can diversify businesses is by doing it ourselves. That way when girls or other minorities look around, they'll see what we've done and feel encouraged. Right now, we are 10–12 people. A lot are freelance contractors and subject matter experts. Before I started, I felt like I had a large network of subject matter experts, the crème de la crème, for any particular energy subject. I can pull people in and use what they know depending on the research we are doing.

We do all kinds of data analysis. We want to tell the story of what the data is saying and come up with meaningful recommendations to accelerate the transition. Our bread and butter is a specific type of analysis called evaluation. We do a different kind than other firms, more agile. We work closely with the companies we evaluate. I would say we do embedded evaluation. We give constant feedback. Writing proposals is one of the most exciting parts of my job. I get to unleash the entrepreneurial thing inside me. I get to feel like I'm competing. I like to think about proposals like I'm on the other side. People don't say exactly what they want, and I have to read between the lines. I love that it's a tight deadline.

One piece that's tough as a woman-owned business is that we are almost never able to win contracts on our own and be the prime contractor. We partner with other companies. I've seen situations where friends have helped win a contract using diversity qualifications, been written into the budget, and then seen the budget cut after winning. I'm careful. I make sure a certain amount of the budget is assigned to us. No one has done this before, so customers don't know they need embedded evaluation. One of the barriers is that companies don't know their performance would be so much better with our involvement, but they still don't want to share the cost. Everyone wants to keep their work a secret or is scared to try new things.

It's a complicated industry. There are intersections between our work and other industries. For example, low-income programs across the country are working on health and safety measures. Many don't know there are energy savings inside their projects they aren't accounting for. A lot of homes in the poorest states don't have air conditioning. People used to be able to make it through summer without AC, but now they can't. Conversations that account for these differences are important for equity. Do we really want people installing room ACs? If they need to cool their entire house instead of one room, it will end up costing a lot more in the long run. I help programs establish what the baseline should be to ensure we help people and the climate and don't penalize low-income households.

If I had a magic wand, I would change cost-effectiveness. I would ensure that energy for every hour of the day is calculated correctly and linked to wholesale markets. We need to manage the grid in a way that makes sure we are measuring and forecasting energy consumption. I think we'll be there in 10 years. I have all of these skills: planning, policy, mathematics. I think a lot of them come from being an immigrant. I have the Eastern way of thinking and the Western way of thinking. The Eastern mentality is more transcendental, the Western more practical and logistical.

It's almost poetic. Half of me figures out details and half has ideas and concepts. I question everything and keep my mind open.

I studied architecture because it merges those two sides. Then I ended up in energy efficiency.

Part V
Building the Grid of Tomorrow

Exciting and terrifying are just different labels for the same thing. The label depends on your team.

These three chapters are about people building infrastructure that doesn't fully exist yet: state EV programs, public fast-charging networks, and microgrids that keep the lights on when the rest of the grid goes dark.

One man works in state government. He helped get $730 million to deploy before 2030 to get more EVs on the road. They testify in Public Utilities Commission cases, and once watched every minute of a five-day hearing because they knew they'd be called. Everything takes more time than you expect.

One runs communications for the largest public fast charging network in the country. She described how their team picks locations, and how they get people interested and engaged with charging. Their CEO likes the Wayne Gretzky line about skating to where the puck's going to be, so the site team picks locations by power supply, then by amenities. Every charger gets a human name, and on Vulcan Street they all got Star Trek names. The goal is to turn range anxiety into range confidence, without making EVs feel like they belong only to privileged west coast liberal elites.

One of them watched Winter Storm Uri light up Texas microgrids while the rest of the grid went dark. Businesses exported power at market rate. People had a place to get light, heat, food, and charge their phones. Most of the country can't export at market rate. That's the gap they're trying to close. They scored all 50 states and came away from a policy summit with about 23 actions state leaders can take. They're workshopping resiliency enterprise zones so a cul-de-sac can jointly invest in shared batteries without it literally being a criminal act.

We just hit 5% EV registration in the United States. By 2030 we might be over 50%. The grid is being built at every layer at once, by people who can hold the work and keep moving.

The EV Program Manager

Pacific Northwest

Part of my work is building the charging network that makes electric vehicles practical for everyone. I work in government, in transportation policy.

My job is to increase adoption of electric vehicles by removing barriers: lack of charging infrastructure, high upfront costs, lack of awareness, and limited model availability. That last one is the toughest because it's largely out of my control. We do a lot of market research. We invest in education and outreach to consumers and fleets.

Right now we're in the middle of stakeholder meetings for a new 10-year plan on EV adoption, including two sessions focused on communities most affected by pollution. We finished an EV equity study with a survey of about 1,500 people. We asked where they get information when buying a car, and which sources they trust.

The answers varied a lot by age and background. YouTube came out as a trusted source in some groups, while for others, car dealers were still number one. One of our biggest victories came with recent legislation putting $730 million over the next decade into transportation electrification. When our programs began, the aim was to assure people that they could do daily driving in an EV. We supported community charging in malls, recreation centers, workplaces.

Over time, we realized people also think about trips they take a couple of times a year, like camping. In 2018, we launched a competitive process for high-speed charging along 33 key transportation corridors.

The difference is huge. And it's not just our work anymore. Private companies, grocery stores, fuel providers, and local governments are all investing in high-speed charging now. I do some policy work in Public Utilities Commission proceedings.

A. Kelly, *Good Work*, Women in Engineering and Science,
https://doi.org/10.1007/978-3-032-27547-9_30

I've testified in several cases: writing testimony, answering discovery questions, preparing for cross-examination.

In one proceeding, I watched every minute of a 5-day hearing because I knew I'd be called to testify and wanted to know exactly what had been said.

In one $110 million plan from a major utility, many of our recommendations were adopted. It was very rewarding to see a stronger, more impactful plan take shape. We can't move as fast as we might like. We launched a major program in 2017 hoping it would be complete by now, and we're only halfway there. Being the first to do something means learning as we go with no examples to follow.

Everything takes more time than you expect. The most impactful part of my work is managing people. My effectiveness is multiplied through my team, who touch so many parts of the market at once.

The hardest part is keeping all the plates spinning. I never get everything done in a day, but I always get the most critical things done. That might mean delegating, postponing, or working longer hours.

The truth is, nothing is ever fully ignored. It's always a juggle, but the important things keep moving.

The Charging Network Manager

West Coast

I am focused on building charging infrastructure across the United States for everyone to use. We're creating a public fast charging network from coast to coast.

Right now we're the largest public network. I'm a senior manager of communications and marketing, focused on customer education.

I teach people about what it's like to use a public fast charger. It's similar to filling up with a gas hose, but not the same. There's more tech involved. You usually have to download an app and sign in. There's a language around it that changes all the time. It all adds up to confusion.

I'm responsible for getting more people to feel confident about charging in public. The types of people I talk to range from dedicated car lovers to people just learning about EVs. The dedicated ones drive electric because they love how cool the cars are, not necessarily for environmental reasons. More like tech innovation.

There are folks who've been driving electric since the first one came out, and those who are brand new, and tech buffs. The average age of an EV driver in America is about 40. That's finally skewing younger as prices come down. I oversee a team who crafts responses to questions or complaints. Maybe a charger not working. Maybe they don't understand how to start a charge.

The work I do is based on the questions people ask, so we can hopefully head them off. We're trying to close that information gap. It's powerful for people to have the same level of knowledge.

We do a lot through our digital platform. We run digital advertising campaigns with talking points that get rolled up to executives when they speak at conferences. It's inevitable, to a point. We still need faster adoption than we have now. The education gap is one reason it's been slow, along with economies of scale for batteries.

A. Kelly, *Good Work*, Women in Engineering and Science,
https://doi.org/10.1007/978-3-032-27547-9_31

It's difficult to develop digestible explanations for a technology-heavy backend. It's like synthesizing science, making clear that people don't need to know *why* this is happening, but rather *that* it is happening.

I measure success by noticing what questions we receive and how frequently. We collect qualitative data too, like going to a charging station and listening to conversations. I'll talk to drivers and approach their actions with curiosity. For now, charging isn't the same as refueling at a gas station. It might take 30–45 min, not a 5-min pit stop. You're supposed to charge while you shop at a grocery store. It requires a shift in mindset where it's not meant to be out of your way.

We have a great site development team that figures out where to put chargers. The biggest factor is power. We need an existing power supply that is big enough. We look at nearby amenities. Where are people going to want to hang out? Is there a restroom, a coffee shop, a shopping center? We build for present and future. We might put a station where 30% of drivers already have EVs, but we also estimate that some areas will hit 30% in 5 years, so we build in anticipation.

Our CEO likes the Wayne Gretzky line: "Skate to where the puck's going to be."

We also want to install for environmental justice and equity reasons. Early adopters tend to be wealthy, but we want charging accessible for everyone. We don't own the land. We create symbiotic relationships with businesses. They host our chargers and get additional customers. Eighty percent of all EV drivers buy something while charging.

One fun aspect is naming chargers. Every single public fast charger has a human name. We look for historical figures or local legends and environmentalists we can honor.

There was a station on Vulcan Street where all the chargers got Star Trek names. We get to have fun with it.

We have ribbon cuttings with site development, customers, policymakers, and local officials like mayors. They know we need everyone: residential and commercial efficiency, renewables, transportation electrification. All of it together. Now demand is coming to us. People ask when we're coming to their town. That's when we knew we were starting to make a difference. Our goal is to flip the historical fear of range anxiety. People have talked about it for years, but it's really baseless. The average driver only goes 40 miles a day.

We want range anxiety to turn into range confidence.

Up and down the west coast, there are charging stations every 50 miles. If you want to drive from one end to the other, you'll be okay. You can do it in an EV. Now that more people are driving EVs, it's like a train of education. People who own them show others and tell them about their experiences. Little circles of people become waterfalls, become cascades.

There's skepticism to overcome. Anytime someone tells you they're doing the right thing, there's distrust. It's a weird time to be alive. Lots of mistrust in institutions and science.

I try to bridge a divide. We don't want to alienate anybody and make them feel like EVs are just for privileged west coast liberal elites.

Part of my job is messaging that's inclusive and not partisan. I can't be pedantic. Instead I use financial arguments: the 80% spending stat, the benefits to local businesses. There was a recent article about the S-Curve of technological adoption. The same thing is happening with EVs that happened with smartphones. Very few people got in at first. Once 5% of the market got in, the adoption curve went vertical.

We just reached 5% EV registration in the United States.

I bet by 2030 we're going to have more than 50% EVs. I'm very optimistic about the long-term perspective. Millennials are starting to surpass baby boomers. People buying their first car will speed this transition.

Check back with me in 4 years, and we'll see if I'm right.

I don't know if I'm an optimist, but EV adoption will absolutely happen.

The Microgrid Consultant

Pacific Northwest

During the big storm in Texas (Uri), there were microgrids at some private facilities. Those businesses stayed up and running even though the entire grid was down. They were able to export power and got paid market rate. They not only provide individual business resilience, they provide community resilience.

That's possible in Texas because of their open market. In most of the country, you can't export your power at market rate.

That's the gap I'm trying to close. I run an advisory and research company. I started it about 10 years ago. The focus is understanding the policy landscape around the electric grid in the United States, what's happening at state public utility commissions, which design the rules around what utilities can do.

Increasingly I kept bumping up against regulatory barriers to innovation and clean energy. I've been motivated to understand what the chinks in the armor are that bring more clean energy into the grid.

Looking at it from a macro level, the utilities industry is one of the most polluting. The North American electric grid is one of the worst. I look for where the biggest problems are. A little over a year ago I partnered with an organization in microgrid journalism. We learned there was a need for a unified voice for the microgrid industry: education about what microgrids are and aren't. Where there might be opportunity to address resilience, energy equity, and climate change. Those are the three pillars our advocacy is built around.

We're trying to be the connective tissue between different groups working on microgrids: conversations between industry, state officials, and federal officials. We want to discuss where microgrids fit in the energy ecosystem and what barriers need

A. Kelly, *Good Work*, Women in Engineering and Science,
https://doi.org/10.1007/978-3-032-27547-9_32

to be overcome. A lot of what we've done is outreach. We go to events where these organizations meet. I do educational briefings, teach about microgrids and what progress they've made.

We outline our work in a vision document with major policy areas that need attention. In general, we're agnostic about specific policy remedies. The electric grid is regulated at the state level and every state is unique. That can be frustrating, but also creates opportunities. You really need to know how South Dakota operates differently than New York, differently than Texas.

We walked through and scored all 50 states, put that together as a report. We engaged policymakers and asked if we got it right. We hosted a policy summit in conjunction with the largest microgrid conference in the United States. We invited federal lawmakers, state officials, and industry representatives. At the end, we came up with about 23 different actions that state leaders can take.

We can help them understand how to establish resiliency across agencies by identifying common needs, for example, the transportation system and the emergency response system. One of the biggest motivations behind this work is that there's real risk. The federal government is investing a lot in infrastructure, and we want to make sure microgrids serve as connective tissue to the underlying generation technologies.

There's a connection between how solar goes into battery storage that fast charges a car that can then feed energy back into the grid during disruption. It takes coordination, control, and intelligence. That's what microgrids offer.

Even during dire conditions, people have a place to get light, heat, food, and charge their phones. That's the gap I'm trying to close.

Part VI
The Efficiency Ecosystem

When you say "efficiency programs" to someone outside the industry, they probably think about light bulbs. Good. Glad the marketing is working.

But there's a whole world behind that. The people in these three chapters design programs, evaluate whether they worked, advocate for better policy, build the software that measures savings, and try not to lose their minds when the same people ask the same damn questions they've been asking for 40 years. These three chapters are a program-design consultant, a judicial-branch portfolio manager, and a Portland engineering firm leader.

One of them spent 20 years in large organizations before going independent. As industry and agriculture sector lead at a state energy trust, they were asked to double savings in 5 years. They invented an incentive that paid a small amount of money for every kWh saved, so internal labor could count. They came up working class. They say there's a hundred of them in the whole country who have any real knowledge of industrial. There's no such thing as a post-industrial society.

One of them works for a state government in the judicial branch, possibly the largest judicial system in the US. Over 400 courthouses. $5 billion in deferred maintenance over 5 years. They work with rural vendors who send a bill on the back of a postcard, alongside investor-owned utilities with efficient systems. When they started in 2019, paper bills got keyed into spreadsheets by hand. Now 85% of those invoices are processed automatically.

One of them runs a 50 to 60 person engineering firm in Portland, mostly mechanical engineers, serving utilities and end-use customers in Oregon and Washington. They subscribe to servant leadership. 11 out of 11 people they have hired are amazing. Moving the needle in the right direction is still important to them. It's not going to be enough forever but it's enough right now.

Energy efficiency is on the front line against climate change. The whole point of the industry is to put itself out of business by getting things to 100% efficiency. By the time we're there, hopefully we'll have the technology we need.

The Industrial Efficiency Engineer

Midwest

I am a consultant. I work with the agencies and utilities that run energy efficiency programs to help them better procure and define the market. I design the programs, I define them, and I optimize them. At this stage, I am mostly working behind the scenes and not interacting directly with the market. I help them answer questions like how to reach energy users, what you're offering them, and how to get that to result in measurable energy savings.

I started my own business 5 years ago and currently work independently. I had worked in very large organizations for 20 years and had significant staff, many meetings, and substantial HR responsibilities. I reached a point where I was in senior management, still focused on programs, and the next step was either becoming an executive and stepping further away from the work, or stepping out entirely and being able to do more of the work myself.

It was really a question of what I wanted to do day to day. I felt like I was going to be stuck where I was for the rest of my career. Go up and out, or do my own thing and do more of the work that I love. I chose B. I've been a program administrator for many years at the federal level and at the state energy trust. I've worked on utility resource acquisition programs and market transformation. Our field is changing all the time. We have moved to decarbonization which switches our focus to electrification. We are interacting more with distributed generation and demand response.

There's a requirement to innovate and change the way you ask people to make programs. You have to change the way you navigate regulatory requirements. People come to me when they have new marching orders and need to know how to do it effectively. My contribution looks like optimization. I help people steer through change and do it creatively. I get my eye on the prize and figure out how to get what

A. Kelly, *Good Work*, Women in Engineering and Science,
https://doi.org/10.1007/978-3-032-27547-9_33

you're after. I used to get made fun of by an employee because I always say "plan and execute, plan and execute." That's where I live.

One of the main things I work on is checking if programs are evaluable by a third party. I will set up formal KPIs because I'm very analytic by nature. A few years ago I asked a CEO of a high-tech biopharma the secret of his success. He said he wanted to monitor what happens and change quickly to adjust to what you learn. Adaptive management. This is what I'm interested in as well.

How do we set up fast feedback mechanisms to know if what we're doing is working? One of the most important things we can do is actually writing things down. That's how you drive change, by systematizing it. I'll talk a little bit about California since they are so complex. The regulator has primary responsibility for making sure change works. California invites in a million stakeholders. There's a lot of public process around rules when they change. When regulators aren't sure if something is going to work, they set up working groups of stakeholders. These folks get together to kick and poke and prod and recommend and find ways to make things work.

On the other hand, there's been so much rulemaking for so many years, and the rules have been founded on an outdated policy model. That model has changed and the rules haven't had a wholesale redo. It's like a ship that gets covered in barnacles. You can paint over the barnacles, but basically there's a bunch of old rules still in place that don't align with the current model.

People are reluctant to let go of structures they decided to work on 10, 15, 20 years ago. There's somebody or somebodies holding that really closely as a full-time job. I'm going to talk about Strategic Energy Management for a while now.

When I got to the state energy trust, I was the industry and agriculture sector lead. We were asked to double energy savings from industrial and agriculture in 5 years. We were told we'd have the budget we needed. The program had equipment measures but didn't have operations and maintenance. Low and no cost savings were not being tapped.

All of a sudden, two thirds of the custom projects in our pipeline got canceled. So never mind doubling savings, we weren't going to be close to our goals. We started doing Strategic Energy Management and realized that we needed some kind of operations and maintenance measure to account for those savings.

We had to change the idea that you could only pay a portion of the customer's cost in an energy efficiency program. In all other projects, we told people that we would pay 50% and the customer would pay 50%. It was based on the philosophy of having skin in the game and that ratepayers can't be paying for all of this.

In this case, I said we may be doing energy savings that have no cost. Maybe internal labor. The ability to track those costs would be really hard. We just created an incentive that was a small amount of money for every kWh the customer saved.

That was a big shift and it's still a barrier in other places. They can't get over the idea that you're not basing your incentive on a portion of project cost. But I believe customers have skin in the game without putting money in the project. They have to invest their time and energy and their own labor and their own thinking. They have to give people the ability to implement projects. I want different skin in the game:

labor and optimization. Strategic Energy Management is totally focused on letting people use their intelligence and skills to manage their buildings' energy use the way they manage everything else at their business.

Everywhere I go, I see that internal labor cannot be counted as a program cost, and I changed that. We use county average labor rates and report it in a form. There's an overt sense that there shouldn't be an incentive for people doing the work themselves and not hiring a contractor. I come from a very working-class background. People that work in our field by and large don't have our background. They are white collar, middle class. They worked in offices and went to school. They don't know about industrial facilities. They think that program design knowledge that works in commercial and residential programs somehow applies to industrial when it doesn't.

For example, programs think people can't install their own lighting. But they're electricians. They can go to a distributor, get a better price, and do the project themselves. Why would you pay a contractor significantly more to do work that your own staff could self-perform?

So SEM was really focused on harnessing technical people that work at these sites and get them working on energy at their sites. That's really different from the normal approach where you want to get the customer out of the picture to whatever degree possible. I think the idea of improvement in climate remains contentious across the country. Addressing the carbon emissions associated with manufacturing is a massive undertaking. Due to politics, the industries themselves will always push toward a technology investment idea. They would like DOE to spend billions of dollars so they can do materials change and make believe that they're all ready to do energy efficiency but it's not true.

I'm scared about it. There's a whole lot to be scared of. I'm not optimistic because the people that I meet, who work in this field, have no connection to industry. Those of us who do all know each other. There's a hundred of us in the whole country who have any real knowledge of industrial. And a lot of people who work in industrial are not liberal lefty climate activists. They're heavy equipment people and they might even be conservative. I think the way you get over it is that people getting interested in this topic need to make the bridge that if you want to do environmental good you need to go to where it's worst. This is not a PR campaign where people who are already doing well should do better in their houses. We have to decarbonize manufacturing and decarbonize industry.

I see a reluctance to go there because of a lack of knowledge. It's going to require people to get out of the mentality that there are villains and there are good guys. That's the big leap that has to happen because we all use things in our day-to-day life. We cannot outsource our solutions to poorer parts of the world. That is not a good model.

We want a nice, clean, and optimized manufacturing industry right here in the United States, with working class wages that support families. That's what we need. We can do it ourselves, and it can be cleaner and better. The solution starts with fixing the biggest industrial emitters. The solution doesn't involve closing those things down. There's no such thing as a post-industrial society.

On my first day of my last job, my supervisor asked me why we should care about any of this with the idea that we were in the post-industrial economy. I was shocked. Oregon is the second most industrial state as a portion of our economy. We are not in a post-industrial economy. To solve this, we need to not demonize people.

The Public Buildings Manager

Pacific Northwest

I work for a state government in the judicial branch. It's probably the largest judicial system in the United States, and potentially worldwide. We have a really large portfolio of courthouses with over 400 buildings and close to 20 million square feet of real estate.

The challenge I work on is that the buildings were previously managed by local entities and have been transferred to the state in the last 20 years. The thinking was that management at a central level would be more effective and that the buildings would receive the same level of funding. But that poses some unique challenges.

My role is to be the driving force behind all sustainability efforts: energy efficiency, conservation, recycling, rolling out renewables. I lead a small data analytics team and subject matter experts on sustainability, energy efficiency, and water conservation. Our program has a significant lack of funding. To give perspective, we have $5 billion in deferred maintenance that needs to be addressed in the next 5 years. That's a big problem because ultimately that's what's contributing to our buildings being inefficient and contributing to climate change. They haven't received the maintenance they need to be efficient and emit less carbon.

I have been in this job for 3 years. One of the major areas of improvement has been improving data quality for energy and water. Our utility spend is like $60 million a year. We manage the utilities and their data for about 170 of the 400 buildings.

The complexity is that the buildings are geographically very spread out. We work with a very large number of vendors with different levels of maturity in their procedures. We work with rural vendors who send a bill on the back of a postcard, then we work with investor-owned utilities that have really efficient systems. We probably have 200 vendors in energy and recycling. When I started in 2019 all of that data

A. Kelly, *Good Work*, Women in Engineering and Science,
https://doi.org/10.1007/978-3-032-27547-9_34

was manually entered. A paper bill would arrive, and someone would key in a couple of the data fields into a spreadsheet from thousands of bills every month. I had never seen anything as antiquated as what my organization was doing when I got here.

The way it works now is that we have migrated 85% of those invoices and they are processed automatically, delivered overnight electronically into our system. We have had a major success with that. Now that we can disseminate monthly energy reports for our colleagues, they can get more familiar with how each building is operating and can minimize their energy. I started into a newer team. The organization had restructured before me and lost everyone. I was the first one into the new team and I was like what is going on? In the public sector, you need to be very patient. Things change slow. It's about bureaucracy and red tape. But if you're in it for the long haul, then you just need to keep at it and continue making improvements.

My predecessors had embarked on this project before I joined. They had started it in 2017 and then given up. The data was very poor quality. I think my assessment was that our colleagues couldn't match the two datasets and they had to give up. I managed to convince my manager that we could work with a software provider and acquire the data through a third party. We managed to save the project and get it out of the grave and resuscitate it. We receive a significant number of ad hoc requests for analysis of ongoing issues as well as evaluating incentives for retrofit projects. We are a small team, but it can vary from anything like figuring out what to do because a homeless person is sleeping in the area for trash pickup at a building, and we need to come up with a solution. We try to get help as much as we can.

In the last year and a half, we've been doing feasibility analysis for renewable energy projects: solar panels and battery storage. Because of my prior experience working with ISO 50001 for energy management, I've been working to put the foundations of a program in place. Now that we have improved data quality, we can set our baseline year for our portfolio for energy, water, and carbon emissions. We're working now on achieving targets for 2030. They aren't formally adopted but have been discussed with different teams. The targets are quite aggressive and achievable. Our baseline for energy and carbon emissions was 2018. We're looking to achieve 40% reduction by 2030 compared to baseline. For carbon reduction, we're looking for a 59% reduction by 2030, based on energy efficiency, rolling out renewable energy projects, and cleaning up the grid.

The carbon target is aligned with the science-based targets for the beyond two-degree pathway. We don't have anything beyond 2030 right now. Hopefully in the meantime we'll get our leadership on board to come up with longer term strategy. That's half the problem. The organization doesn't stand behind them yet. Hopefully the management will come in the next 12 months. The thing I'm proudest of is that I've been able to develop a very close knit and effective team which is able to do a lot without a lot of funding.

I was the first person to onboard a person fully remotely in the organization during the pandemic. I then onboarded three people during the pandemic, and two of them have moved back to the private sector.

It's difficult to recruit for the public sector because there's really good opportunities for sustainability professionals out there. Some people lose patience.

In the public sector, you need to be very patient. Things change slow. It's about bureaucracy and red tape. But if you're in it for the long haul, then you just need to keep at it and continue making improvements.

The Energy Program Leader

Pacific Northwest

My company is a technically focused organization with a strong history of energy engineering. Most of our staff are mechanical engineers. Another good portion are technically capable project and program managers. We deliver customized energy efficiency, renewable energy, and distributed energy solutions to our clients.

Our clients typically are gas and electric utilities, but there are also end-use customers: businesses, small-to-large commercial buildings, and industrial customers. We work in buildings and facilities to determine why systems aren't functioning properly or help identify opportunities for efficiency. We work with utility clients to design and deploy energy programs to help their customers save energy.

We're between 50 and 60 employees mostly on the west coast. Our office is in the Pacific Northwest. Our clients are mainly in Oregon and Washington. I manage a mix of energy engineers and non-engineer project managers. The core work in our office currently is strategic energy management, which requires a mix of technical and non-technical skills. For SEM, we consider energy coaches that deliver curriculum through a utility program.

I don't subscribe to any one leadership style, except maybe servant leadership. I took a leadership and management training course about a year and a half ago and have integrated some aspects of it into my work. My style has adapted over the years. In a previous role in a different company, I had a really hard time as a manager for a number of different reasons, and I didn't ever want to have to revisit those situations ever again.

Coming to this company and growing slowly, starting with basically no direct reports and then growing to having several, has enabled me to develop my own style. Servant leadership resonates with me because I may be very motivated in

A. Kelly, *Good Work*, Women in Engineering and Science,
https://doi.org/10.1007/978-3-032-27547-9_35

some ways, but I'm not a super type A outwardly driven individual. I take my time on making decisions. I take my time and get to know people before I make decisions. Servant leadership is about meeting people where they are and empowering people to make certain decisions, guiding them, and allowing them to propel themselves forward. It's about making sure they feel supported along the way, letting them know that I'm there to support them. They can come to me at any point, but I am empowering them to chart their own course as well, just like I was empowered to do.

It's been super successful. I have a very different relationship with each of my team members.

If there's a flaw in my style, it could be the perception of uneven treatment or favoritism. I don't think the team sees it that way. They know that I have personalized relationships with everyone else. Part of the story behind servant leadership is I've built an amazing team. I've hired some amazing people. I look around and go, how did that happen? I feel like I have a certain knack for getting to know people and knowing that they're going to be the right people to be on the team.

At first I thought, is this a fluke? The first couple of employees were such great people. How am I going to keep this going? So now 11 out of 11 people hired are all amazing employees. Because of that I have very few performance issues.

In the few cases that I have, it's a challenge to get to the heart of the issue, but it's about understanding each person's motivations. If there's a performance issue, it's partly my failure. I've brought that person on and have tried to develop them. It takes time. Beyond personnel management and teambuilding, I have applied work functions. I bring design concepts for the programs we deliver. Because we're a small team, I wear multiple hats. I have a sales and business development role where I am developing relationships and identifying opportunities to grow our practice areas.

I am also the de facto marketing director here. I oversee web development and other marketing-related material: collateral, social media. My background and interests in undergrad were architecture and urban planning. I focused my studies on community development from a design and functional standpoint. I realized I wanted to broaden that focus so I decided to go back to grad school and pursue a degree in international affairs and sustainable development.

After graduation, I took an opportunity to work in the Middle East where I became a green building consultant in Dubai. While the work wasn't squarely in my academic area of focus, it was adjacent. At the time, I was a certified LEED accredited professional. We worked throughout the region on a lot of projects, so I was able to travel the Gulf region quite a bit. During that time, I became more interested in the energy efficiency portions of the LEED rating system and understanding the energy using systems in buildings.

There just seemed to be more exciting opportunities from a career path standpoint. So I used that and moving back to the states, helped a local non-profit develop a commercial green building program with a focus on energy. I think what drives me as a professional is the motivation to serve. My previous career as an urban planner was all about serving the public and being in public service. Both my parents were public servants. My mom was a career elementary school teacher, and my dad was a civil engineer working for municipal government. I think I've taken it a step further and recognized a larger call to serve the global community rather than just a local community.

At my company, I am confident we're doing the good work. Yes, I am in a business trying to earn profits and do things for its employees. But I want to make sure it's always for the right reasons. When it comes down to it, energy efficiency and helping people, organizations, communities save energy and reduce greenhouse gas emissions is what I consider purposeful work. My philosophy has changed over time. As a student, it's really easy to be passionate and idealistic that we just need to stop climate change now. Why didn't we listen to all the smart people from the 60 s and 70 s who told us about human-driven climate change.

I feel like my approach now is a bit more practical and realistic about what an individual can accomplish. I'm not a single person. I'm married with two kids and that changes everything. My perspective has changed. I am more focused on family and the future of my children. While I want to create a legacy for the future generations, I also must stay present.

Being able to enrich a child's life through parenthood is super critical and requires being present as a parent. I didn't realize how much influence you have as a parent until I became one. It matters. You can really alter the life of someone really quick.

Moving the needle in the right direction is still important to me. It's not going to be enough forever but it's enough right now. We are doing well and doing good work. I would give credit to my company too. We aren't turning profits for investors. We are employee focused. I have seen firsthand what this company does for its people, like making big changes to healthcare plans to support personal situations. We rally to help people. It's a bit cheesy to say, but we're like a big family.

The last thing I would say is that it's important to understand the underlying science of the climate crisis we face. Issues get confused when arguments about climate change are politicized. Nevertheless, if you understand even just the basics of climate science, you'll carry that with you. I think the decisions you'll end up making in your career or in your life will tend to be positive.

Globally, the world is going to look very different in short order. We just want to be able to live fruitful lives and provide good lives for our offspring and their offspring. I think there are ways to do positive work no matter what path you take in life.

Part VII
The Long Haul

Being able to teach and lead feels like something you can do after you already have a lot of knowledge. It's something you progress to, not a place where you would normally start.

That's why this part is last.

If you've made it through all 38 chapters, you've met the installers, researchers, advocates, founders, grid builders, and the efficiency workers. You've seen what the work looks like across the whole industry. Now you meet the people who carry it forward.

The wind turbine technician started as a bartender. She went back to school, learned about power, and almost didn't go to the second interview because she thought they'd never hire her. Then she climbed a turbine for the first time, terrified, telling herself she was doing it for women everywhere. Now she teaches others to use the software systems that keep renewable sites running smoothly.

The utility policy manager had an undergraduate class with Jared Diamond. He was working on a major book about civilizations collapsing from climate. She realized: "This is the thing, this is my thing!" Now she's balancing climate work with two young kids and wondering how much one person can move the needle.

The ESG consultant wakes up, walks his dog, goes to the gym, and then sits down to help utilities track their carbon emissions. The most important part of his job is teaching, coaching analysts to see the bigger picture, to understand why any of this matters.

These three represent something I noticed across all 38 interviews: the people doing this work don't think of themselves as heroes. They're just trying to help. They teach because they learned. They lead because someone has to. And they worry, constantly, about whether they're doing enough.

What do I want you to think as you close this book? I want you to think about career pathways to get in, or degree programs to get the skills you need. I want you to think about teachers who made you feel smart and important.

And I want you to feel good about the work you already do. Even if it's staring at a glowing rectangle all day. Even if it feels like you're failing.

The Wind Turbine Technician

Midwest

I used to be a bartender. I was a bartender and restaurant manager for years. Then I got to the age where I thought I should get a real job, so I went back to school and started studying solar.

Before long, I saw on the news that a local college was providing free training to people who were unemployed or underemployed. I went there instead. I spent about 10 months learning about power.

I was doing solar at first, then the principal came to me and asked if I wanted to do wind. I said, "No! I think those people are crazy! They climb up high on those turbines and work in a tiny little box. I'm not interested."

But then he told me a company was doing an internship and I could win some money. Then I got interested.

I did the wind class. I got to the interview, and I was thinking it was all guys; the turbines and parts are huge. I thought, "Oh gosh, they're never going to pick me. I'm not even a big girl!"

I almost didn't bother going to the second interview. But I thought I should, so I did.

I got the internship and worked on three different turbine sites. One was 60 one-megawatt turbines, 212 feet tall, with no climb assists. That means you just climb straight up the ladder. You're tied off, but there's nothing to help you. This was 2011. You need to remember I was the only girl working out in the field.

I remember my climb test vividly. I got halfway up the 212-foot tower. I looked up and saw I still had so far to go. Then I looked down and it was so far down.

I remember thinking to myself, "Why am I doing this? This is crazy!"

A. Kelly, *Good Work*, Women in Engineering and Science,
https://doi.org/10.1007/978-3-032-27547-9_36

And then I remember saying that I was doing it for women everywhere. And that I had to do it. I could not fail at this.

So I decided I wasn't going to look up anymore. I would just look straight ahead.

I looked straight ahead and kept going. When I got to the first section and was able to walk around, I stretched and swung my arms in circles. I had been holding on to that ladder so tightly, and my arms were so tired I could barely hold on anymore.

When I got to every level, I would stop and spin my arms around to try to get some more life in them. I made it to the top. And it was amazing. I opened the hatch and poked my head out and looked around.

Being up there is amazing. You can see everywhere all around you. I love climbing a turbine. Especially the smaller, older ones that you open up. You're safe, tied off, hard hat on, and you can be alone. Just out there in the field and there's a nice breeze. You can see for miles, right down into the mountains.

One of the best parts is rescue training, where you practice rescuing a dummy off the side of a really tall turbine. We have to do it every year because honestly, if something happens, you might not get a firefighter or paramedic who can climb the tower. They'll wait at the bottom, so everyone needs to know how to do it. After the internship, I had a normal O&M job. In the morning, we'd see which turbines had faulted. Turbines will shut themselves down in some situations so they don't get damaged.

One time I was driving through the Mojave desert with a friend, and the turbines weren't spinning. My friend asked, "Why aren't the turbines working? It's so windy out!" I told her it was a high wind shutdown. When winds are too high, it'll damage the turbines. They'll change the shape of the blades so they don't damage themselves.

I was out in the field for 3 months. Then I got called into the office and applied for supervisor training. After a few months, I went to advanced wind technician training. I came back and became supervisor for about 350 wind turbines. My job was to make sure everyone stretched every day and all the maintenance got done.

At first, some of the guys didn't like working for a woman and a few quit. But the rest were fine.

We'd go outside and do stretching exercises every day. Then I'd get to work an hour before everyone else to figure out what needed to be done. I'd assign high-voltage work to specialists, send someone out to reset turbines, and send people who were good at troubleshooting to do repairs. Then it was time to decommission the turbines because they got old. My area manager was worried about my position.

Luckily, the whole time I'd been managing the site, I'd also been using our company's software system, and I'd gotten pretty good at it. I found that I really loved it.

It's funny because I'd never really worked on computers. Since I was getting into that software, the area manager decided that I would become a planner and scheduler.

That work meant making sure everything that needed to be done had the right equipment at the right time. If we needed to take a turbine apart for repairs, we'd need a huge crane, a crawler, to come on site. It comes on 12 or 13 trucks and is assembled on site by smaller cranes. Then the crawler takes the whole rotor and three blades to the ground so they can be worked on. After that, a reorganization made me into a business coordinator. I'd become a subject matter expert in the software. Now I work with solar and don't even work with wind anymore. I'm in the office, not in the field.

I also teach people how to use it. I'm like an all-around person that makes sure everything runs smoothly. I have 11 sites. For each one, I check that they have all the parts needed for maintenance. I like to teach. I like to share my knowledge. I've taken way more courses than most people. I think 15. Most people who use the software don't like it and do a bad job. I take as many classes as I want.

It's rare, but if I find someone who likes the software and wants to learn, I'll teach them and continually reach out to show them new things.

I host a class once a month. Anyone can join and ask whatever they want. I just go and they tell me what they want, and I teach them.

I'm going down to one region where we have a lot of solar and new sites and battery storage. The guys are new (have been with the company maybe a year), and there are lots to teach. I really had fun there last year. The guys were really nice. They put up a big monitor, and everyone could watch my work.

A lot of the guys are really interested in learning. It makes them look good. They're very interested, or they're really good at pretending to be interested. People are always saying how smooth our area is. I like it like that.

I have a bit of a trick. You clock in the morning, clock out when you leave. Every 2 weeks, I run a report and put everyone in a table, and they see they haven't entered their time correctly. It creates a sense of pressure. Nobody wants to be the one that let the site down. The company is really good to people. After a year, 5 years, and 10 years, you get a gift. For my ten-year gift, I got a huge kayak, like $900. I might have spent $300 on a little one in normal life, but I wouldn't buy a $900 one.

Another thing we do is give gifts to children in the community at Christmastime. Every year I go to the dollar store and spend $100 and give them away. It's a very poor community. People would be so happy to come and pick up a mug. They'd change their minds over and over and switch out their presents.

I also get to turn on the Christmas lights with Santa every year. That is so fun.

We have lots of people who've been here since the beginning. You can tell it's a good place to work because everyone has been here so long.

The Utility Policy Manager

Pacific Northwest

I had an undergraduate class with Jared Diamond. He was working on a major book. In class every week, we covered a chapter. A key narrative was civilizations collapsing due to climate events. And then I realized, "Oh my God, this is the thing, this is my thing! This is what's going to be my generation's big issue." Ever since that one class, I knew that climate change was my thing. That's what I'm going to dedicate my life to. My utility serves several cities in the Pacific Northwest. I joined about a year ago. Before that, I worked for a federal power administration for 11 years. I work in government affairs as an environmental policy manager. My areas of focus are building decarbonization and energy efficiency. For the most part, decarbonization is electrification to reduce GHGs. We also do this for health equity and indoor air quality.

I mostly focus on stakeholder development and policy development. I just got put on a legislative task force to evaluate policies that legislatures could use next year to drive decarbonization and reduce energy burden. I am really trying to drive GHG out of buildings. One of the exciting things underway with energy efficiency is driving it to focus on GHG reductions. In the past, it was based on the idea that every unit of energy was equal. Now the units that save the most GHG should be sought after. We are trying to do that because it's consistent with our company goals to dramatically reduce GHG emissions associated with the energy we supply to customers. Matching our investments to energy efficiency is also consistent with state law to decrease GHG over time. I'm on the government affairs team, and everyone has a different area of focus. My team focuses on laws related to utilities. When we were in the legislative session, our folks were trying to pass bills that are

A. Kelly, *Good Work*, Women in Engineering and Science,
https://doi.org/10.1007/978-3-032-27547-9_37

beneficial to the utility and to customers and be consistent with our strategy. We get close to the legislative action that happens in Salem.

A big part of my job is managing relations with external parties who have a shared interest in moving the state forward with climate change mitigation. It's a lot of environmental organizations. I stay connected with people and their strategies. When we have areas of shared interest, we can push together for a single outcome.

In 2021, we got a state law passed that requires the two investor-owned utilities to get an 80% reduction by 2030. That was years in the making, and I joined the team in the middle of the session. Now we're looking at what policy might look like in 2023. It's the middle of the brainstorming phase, looking at what actual laws are passed in other states, what laws we've recently passed, and trying to move the needle. Each state has its own unique laws and regulations. In this state, we have electric utilities and we have gas utilities, and that's really different than utilities that have both. In energy efficiency, it is difficult to financially support a customer who wants to switch from gas to electric. If you want to make progress with climate change, those things need to be looked at, and it's hard to move away from them because they've been around for a long time.

The regulatory work is circling around blockers and friction points, identifying opportunities. Not everyone is super motivated by climate change, but they might be interested in dealing with the housing issue, so we'll look for synergies. There's an overlap in interest between those who need more housing and those who want super-efficient and resilient homes. There were so many important bills that passed last year. There was a real thrill and there were big things moving. For me, when my job started, the buildings policy work was a new position that was created. Some of my job has been making that position my own and figuring out what the needs are. It's been fun; I really like to chart my own course.

When I joined, induction cooking was still super underground and you never saw it. But this year, now there's this task force with an even broader scope. We should look at all policies related to all buildings, all sectors, new and existing. It felt like we went from 0 to 60 really fast. The other part of my job is products and services. There's a new program focused on heat pump water heaters that is a big opportunity. There are a bunch of gas hot water heaters out there. How can the utility play a role in making it easier for customers to switch? There are certain things that have to happen in the marketplace. We have to get more heat pumps out there.

There are a lot of barriers. We've got contractors that are totally skeptical about heat pumps. There are higher costs even though they are super-efficient. Each of those little things adds up to a lot of work. It does feel like a lot. There's way too much to do. I actually get nervous sometimes about us trying to do too much to the point where we don't do anything well. The company has a lot of new people coming in. There's so much demand for workers in the energy space because we've had so much retirement. But the new people are all very similar to me. They're super motivated about climate change and have been for a long time. They get in here and want to do super good, super cool stuff. It's awesome. There's so much good energy. But we can't do everything at once. We have limited funds and limited time.

Right now, it feels a little bit chaotic. It's fun to be there where it all feels new, instead of joining the utility 20 years ago when it was the same old vertically integrated utility. There's such a different ball game now. How do I feel about the impact I'm having? That's on the verge of an existential question right there. That's where we get into a very personal space. It's the midlife stuff. You have a family, you have kids. And you've been a person that is super motivated to change the world since you were 10 years old. You've always been driven to do really well at school, go to grad school, and find a job. Then it's decades worth of thinking through priorities and thinking about if you're changing the world.

The change-the-world attitude is so anxiety and stress-inducing. You never really actually change the world. You can have incremental influence on it. You set the bar so high, and if you're not careful on this never-ending treadmill, you'll burn out. It's a total global crisis. It's climate change; let's be real here.

So there are decades worth of self-reflection, talking with friends and colleagues, trying to make progress and move the needle. Also recognizing that you're just one person and you're doing the best you can.

I have two young kids that I love and adore so much, and it wasn't easy to realize that I'm not interested in putting in 80–100 hours of work to change the world. I want to do a super solid job on the time that I'm working, but then I want to be in a good spot to spend time with my family. I think that balance is super important. I worked in public power for 11 years before and was not that excited about utilities. I had been looking at switching for a couple of years, thinking that this utility seemed to have the right stuff and was saying the right things. Leadership here is really serious about mitigating climate change.

It's been super interesting to figure out how to make a profit while doing the right thing. What do we need for the climate, what do customers want, and figuring out how to piece those three things together. We are trying to figure out how to maximize benefits for everybody.

I'm excited to be at a utility that isn't trying to slow responses to climate change. We are cleaning up our electricity supply, our transportation, and electrifying buildings. We are ultimately going to deliver that in a capitalistic ecosystem and show that you can please investors, do the right thing for customers, and mitigate climate change.

The ESG Consultant

Southeast

I work from home, which is nice. I wake up, I walk my dog, I go to the gym, and then I sit down in my home in San Diego and get to work. I check my email and start connecting with colleagues on the project we're working on. Right now, we are working on a large project trying to help California utilities figure out how to collect and manage their sustainability data: carbon emissions, energy usage, water, wastewater. Utilities need to track their performance on sustainability metrics and use those insights to know what to do next. The climate risk profiles that people report on are what comprise ESG. The main components I work on are:

Emissions intensity of running the business.
Emissions impact is based on consumption of the products or services the company provides.
Emissions from creating products, including emissions generated for the raw materials and shipping them around the world.

Climate risk profiles also include water consumption, waste, and hazardous materials. But there's also the social dimension: the ways businesses are operating that contribute to wealth inequality or create upward social mobility barriers. Then there's governance: How safe are your operations? What are you doing to make sure you're not going to create a wildfire that displaces millions of people? How do you consider the physical, mental, and emotional safety of your employees? I'm new at the job, so I only have one big project. There are elements within the project that are small projects themselves. I spend time trying to understand how other organizations have approached similar challenges. I look at how Fortune 500 companies

A. Kelly, *Good Work*, Women in Engineering and Science,
https://doi.org/10.1007/978-3-032-27547-9_38

have organized their sustainability data. I search for patterns, best practices, and things to avoid.

I gather case studies and distill them so I can make fair comparisons. I'll call people doing projects in similar areas. I pull consultant reports and external databases. There are web scrapers that collect publicly available data from financial filings with the SEC. A lot of companies are voluntarily disclosing their climate data. With all this data, I benchmark and compare how each company is doing on their ESG scores. ESG data comes from all over an organization. I have to help folks find fuel receipts that their truck drivers are turning in, lists of suppliers and customers. There are all different kinds of data. The data comes in many forms, often poorly organized or hidden in old spreadsheets. I help them figure out what tools are available and what methods have worked for other large, complex organizations.

But the most important thing I do at work is teaching. I have a lot of consultants and analysts working under me on this project. Coaching and teaching are important to me. I try to make sure throughout the week that I'm taking time to coach my team and make sure they understand what we're talking about and, more importantly, why we're talking about it. I want to keep them connected to a purpose beyond delivering the project and having a happy client. This matters to the world. It's kind of jokey and cheesy, but when you're bogged down in daily details, it helps to remember the bigger vision. There are things people should avoid. First: don't expect a single technology to solve everything. It's tempting to chase the latest software, hoping it will fix everything, but that rarely works.

Software is just one piece of the puzzle. But the biggest barrier, problem, component is the people. You have to align people across their organization so they can see why this is important and why they might need to change their habits or behavior. Getting thousands of employees to follow the same procedures is hard. It's even harder when the connection between their actions and the company's sustainability goals isn't obvious. Building buy-in means meeting with each stakeholder and showing them the bigger picture. I have to make sure that people working on a piece can see the whole. Right now, I'm working with environmental health and safety managers who handle chemicals and hazardous materials. They voluntarily hold a lot of the data we need on water and waste and emissions. But they don't want to participate in putting their data in a new system. They want to design their own systems. They don't see the relationships between what they do and the broader sustainability goals of their business.

There are competing interests, competing visions, and lots of folks looking for a technological savior. The process is slow and starts with culture and strategy, not with hardware and software. Getting buy-in is hard, but the most vital technique is just to listen. You have to be patient. When I talk about incremental progress, the most important incremental progress I have made is in building relationships. It takes patience and it means meeting them where they are. As cliché as that sounds, it really is essential.

We are starting to get indicators of success. These groups are willing to listen to each other and align their vision across different business functions. We create spaces for people to talk to each other. I lead design thinking sessions for brainstorming with different departments. We moderate. We help start the conversations, and soon people share their own stories. I have to communicate with climate skeptics on a regular basis. I encounter more climate skeptics back home in Kentucky with my high school friends, neighborhood buddies, and family members than anywhere else.

When I encounter climate skeptics professionally, I try to think about their role, what reasons they might have, and how their perspective facilitates their work. It takes a long time to change people's minds and deeply held beliefs. When someone's beliefs aren't based on logic, I rarely try to argue.

Sometimes I use a different approach. I'll ask questions rather than speak. I'll ask them what they think about risks of climate change. I'll say that even if there's a 50–50 chance things could get really bad, what would they do to combat the risk? If there was a 50–50 chance your house was going to burn down, wouldn't you move to reduce that risk?

I try to get people to think probabilistically and not in absolutes. I also point out the massive boom in employment that clean energy jobs have brought to our workforce. I try to make it more patriotic. Those tactics do work. I don't try to put my vision of the future onto anybody else. I think it's important that everybody has their own perspective on what a sustainable future world looks like. It's more about finding common threads among those visions to bring people together.

For me, it's about generational equity. That's why I made it my life's work. I think that the world around us is something that we share. We borrow and give to each other over time. I want to make sure that the next generation of people has the same safe and abundant world that I had growing up. I want them to be able to go and hike the same hikes that I hiked and canoe the same rivers and creeks that I canoed in my home state of Kentucky. I recently changed careers because I wanted to have a bigger impact. At my old role, I was doing work that felt like spinning my wheels. I had accidentally worked myself into an administrative role and less of a role in getting things done.

In my current job, my focus is on getting things done. Little by little, I think through all the things I'm doing on a daily or weekly basis. I try to keep the ripple effects of my work in my mind. That helps me feel like the impact is there.

I like to think about how my work, combined with the efforts of thousands of other climate workers, might add up over time. You have to look incrementally because there are very few home runs. In climate, it's always two steps forward, one step back. You just have to keep hacking away.

Appendices

Appendix: Career Resources

Resources mentioned by interviewees throughout this book

Entry-Level Pathways

Direct Entry (No Degree Required)

Solar Installation - Start as crew member on residential or commercial installations—Company training provided; prior electrical experience helpful—Military veterans often transition successfully—"If you're one of these people who had done Lego robotics since elementary school, great. But if you just got good at math, that works too."

Wind Turbine Technician - Many community colleges offer wind technician training programs—Free/subsidized training available for unemployed or underemployed workers—Internship programs with on-site experience—Can advance to supervisor, then office/planning roles.

Crane Operator - Start as "yard kid" learning rigging and components—Apprenticeship through local union of operating engineers—"Crane operators make a couple hundred grand a year if they're nice to work with."

Utility Field Technician - Entry-level positions in solar plant O&M, starting around $12/h—Advancement to site manager possible within 2 years given high demand—"If you want to be, you can start out as a high-performer and theoretically be a site manager in two years."

A. Kelly, *Good Work*, Women in Engineering and Science,
https://doi.org/10.1007/978-3-032-27547-9

Technical Certifications

LEED Accredited Professional (AP) - Leadership in Energy and Environmental Design certification—Demonstrates competency in green building practices—Useful for consultants, architects, engineers in sustainability

ISO 50001 - Energy management systems certification—Used by facilities managers and energy program leads

ENERGY STAR Certification Programs - Various certifications for products and buildings—Relevant for market transformation and efficiency work

Degree Programs and Academic Pathways

Undergraduate

- Engineering (mechanical, electrical, civil)
- Environmental science
- Architecture → green building specialization
- Psychology/social science → program evaluation and behavior research
- "I got into engineering because I was good at math. I was not one of these people who had done Lego robotics since elementary school."

Graduate

- Energy policy and management
- Environmental policy
- Urban planning with sustainability focus
- International affairs and sustainable development
- "I went back to grad school and pursued a degree in international affairs and sustainable development."

PhD Pathways

- Environmental microbiology (climate research)
- Pure/applied mathematics (data science, machine learning)
- Geological engineering (mining, resource management)
- PhD students: National labs offer summer research programs for first-generation students

Government and Federal Opportunities

National Laboratories - Summer research experiences for undergraduates—Programs targeting first-generation college students—Network access to researchers and practitioners

Department of Energy (DOE) - ARPA-E funding for innovative energy companies—Grant programs for clean energy development—"We were blessed to get non-diluted government funding in the beginning from ARPA-E."

State Energy Offices - Policy analyst positions—Program management roles—Regulatory affairs

Public Utilities Commissions - Regulatory proceedings and stakeholder engagement—Policy development roles

Nonprofit and Advocacy Organizations

Environmental Advocacy - Sierra Club—Natural Resources Defense Council (NRDC)—Regional climate justice organizations

Workforce Development - A workforce nonprofit: training next-generation green workforce—Internships and in-classroom programs for high school students

Trade Associations - Energy efficiency implementer associations—Solar Energy Industries Association—American Wind Energy Association

Private Sector Entry Points

Utilities

- Customer programs (energy efficiency, demand response)
- Environmental policy departments
- Community outreach positions
- "I've been at the utility for 3 years… It's been super interesting to figure out how to make a profit while doing the right thing."

Consulting

- Energy efficiency consulting firms
- ESG (Environmental, Social, Governance) consulting
- Program design and evaluation
- Technical engineering consulting
- "Consulting is pulling information from different sources to give people a starting point, then taking critique and making it better."

Clean Energy Companies

- Solar developers and installers
- Wind farm developers and operators
- EV charging network companies
- Energy storage companies
- Microgrid developers

Startups

- Hardware companies (methane capture, battery systems)
- Software/SaaS (energy management, grid optimization)
- Clean tech consulting
- "We have bootstrapped this business from the beginning."

Job Boards and Communities

- Climatebase (climatebase.org): Climate-focused job board
- Work on Climate: Slack community with job channels
- Climate Draft: Matches talent to climate companies
- IREC Job Board: Interstate Renewable Energy Council listings
- Terra.do: Climate career community and courses

Skills Employers Mention

Technical

- Electrical troubleshooting and multimeter use
- Data analysis and visualization
- Energy modeling
- Machine learning/data science
- "Most of what I do is reviewing, interfacing with clients, making decisions, pointing people in the right direction."

Non-Technical

- Public speaking and presentations
- Persuasive writing
- Stakeholder engagement
- Project management
- "Persuasive writing is important. It doesn't come naturally but it has to cook in the brain for a while."

Soft Skills

- Teamwork ("What I really need is somebody who played sports." (Ch13))
- Listening and curiosity
- Patience with bureaucracy
- Adaptability

Salary Information

A note on compensation: Salary ranges in this appendix reflect what interviewees shared during 2022–2023 interviews and may not reflect current market rates. For up-to-date compensation data, consult the Bureau of Labor Statistics Occupational Outlook Handbook (bls.gov/ooh) or O*NET OnLine (onetonline.org).

Role	Range/Notes
Crane operator	"A couple hundred grand a year" (experienced)
Solar plant technician	Entry: ~$12/h; advances to "good quality of life"
Wind turbine technician	Good stability, especially in remote areas
Energy consulting	Variable; can bootstrap to independence
Utility positions	Stable with benefits; government scale

Note: Salaries vary significantly by location, experience, and employer. Interviewees spoke in general terms about compensation.

How to Get Started: Advice from Interviewees

Network actively: "The lesson: fight your fear. This is a collaborative industry."

Say yes to opportunities: "My approach: say yes and then figure out how to do the work."

Take daily action: "Every day I woke up and took at least three steps to further my career. Whether that was finding new connections, even if I didn't feel like it."

Get experience anywhere: "If they had a high school class on Bolts and Volts with a bit of computer networking, it would be killer."

Write and share: "I started a blog to answer questions I found interesting. It snowballed into a consulting practice."

Build diverse skills: "You have to be a good listener. People have stories to tell. When you meet someone with a story, sit and listen."

Key Takeaway

"The coolest thing about climate change is that there is space for absolutely everybody. Everything impacts climate change. Whether you're in the restaurant industry thinking about sustainable food production, in marketing wanting to work for a clean technology startup, in a factory figuring out more efficient ways to run, there's space for everybody. Everybody has a role to play in the climate fight." *Resources compiled from interviews conducted 2020–2024. For current job listings and program information, search organization names directly.*

Appendix: Sources

The following sources support key statistics cited in this book:

Clean Energy Workforce - International Energy Agency, *World Energy Employment 2023*: Projects 14 million clean energy jobs globally by 2030.

Buildings and Emissions - UN Environment Programme, *2022 Global Status Report for Buildings and Construction*: Buildings account for approximately 40% of global energy-related CO_2 emissions.

Cooling Demand - International Energy Agency, *The Future of Cooling* (2018): Global cooling demand is growing faster than any other end use in buildings.

Battery Technology - U.S. Department of Energy, Office of Energy Efficiency and Renewable Energy, *Battery Basics* and industry data: Standard lithium-ion EV batteries typically achieve 1,500–2,500 charge cycles; lithium iron phosphate (LFP) batteries can achieve 6,000–10,000 cycles.

Electric Vehicle Adoption - U.S. Department of Energy, Alternative Fuels Data Center: EV registration reached 5% of new vehicle sales in the United States.

Critical Minerals - International Energy Agency, *The Role of Critical Minerals in Clean Energy Transitions* (2021): Clean energy technologies require significantly more minerals than current recycling can supply.

Appendix: Methodology and References and Sources

How This Book was Made

I first read Studs Terkel for a sociology class in college. Working people talking about their jobs, janitors, steelworkers, waitresses, with dignity and specificity and the mess of real life. I wanted to do that. Not the same jobs, but the same thing: let people say how they did it, in their own words.

I picked climate because I wanted to show people they could have a career with meaning. Not just scientists. Not just engineers. Everyone. If you're a crane operator, there's climate work. If you're in sales, there's climate work. If you're a policy wonk or a community organizer or someone who just wants to feel good going to bed at night, there's climate work.

I interviewed 50 people about their work in climate. Most conversations happened over video calls. Not exactly the stuff of gritty documentary.

But what I heard was real. People told me about jobs they'd stumbled into, mentors who changed their paths, days they wanted to quit, and moments they realized the work mattered. They told me about divorce and dead parents and kids who grabbed their briefcases and cried. They told me about exploding wind turbines and oil spills and the small daily grind of making bureaucracies move.

The Problem with Transcripts

Raw transcripts are unreadable. People don't talk in paragraphs. They circle back, restart, and contradict themselves. That's what makes speech *alive*, but it's also what makes it hard to read.

The traditional approach is to clean it up. Fix the grammar, organize the thoughts, and make it flow. But every edit moves you further from the person who actually spoke. Polish too much and you lose the voice. Don't polish enough and no one can follow it.

I needed some kind of framework for deciding what to keep and what to cut.

Ten Readers, One Story

I started thinking about who would actually read this book. Not one reader, many, each with their own questions.

A college student wondering what climate careers exist. An activist wanting to understand how change happens. A mid-career professional looking for permission to switch fields. A skeptic checking whether these stories hold up. An editor asking whether every sentence earns its place.

I turned these readers into personas; imaginary reviewers each see different things.

The Ethnographer: Does this sound like a real human? Is the voice distinctive?

The Student: Can I picture myself in this job?

The Skeptic: Does anything sound made up?

The Heart: Where's the emotional core?

The Career Seeker: How did they start? What skills matter?

The Climate Reader: Why does this work matter?

The Editor: What's redundant?

The Brevity Reader: Could this be shorter?

The Opener: Would you keep reading after the first paragraph?

The Closer: Does the ending land?

The Process

Each chapter went through these ten lenses, multiple times. Where three or more personas agreed something needed to change, I changed it. Where they disagreed, I made a judgment call.

But here's the rule I never broke: I didn't add anything.

Everything in these chapters came from the interviews. I could cut, rearrange, and add context from elsewhere in the same conversation. But I couldn't put words in anyone's mouth. These are their stories, not my reconstructions of them.

The goal wasn't to make everyone sound polished. Some of these people are eloquent. Some ramble. Some get mad. Some cry. I kept the texture, the mess of actual humans talking about work they actually do.

A Note on References

This book is a collection of oral history interviews in the documentary tradition of Studs Terkel. Unlike traditional academic texts, the content comes primarily from first-person accounts rather than secondary sources. The interviewees speak in their own words about their experiences, observations, and perspectives.

References in this book are therefore limited to external statistics and facts cited within the narrative, rather than chapter-by-chapter bibliographies. The sources below support the key data points mentioned across chapters.

Sources

Clean Energy Workforce

International Energy Agency. *World Energy Employment 2023*. Paris: IEA, 2023.

- Projects 14 million clean energy jobs globally by 2030.

Buildings and Emissions

UN Environment Programme. *2022 Global Status Report for Buildings and Construction*. Nairobi: UNEP, 2022.

- Buildings account for approximately 40% of global energy-related CO_2 emissions.

Cooling Demand

International Energy Agency. *The Future of Cooling: Opportunities for Energy-Efficient Air Conditioning*. Paris: IEA, 2018.

- Global cooling demand is growing faster than any other end use in buildings.

Battery Technology

U.S. Department of Energy, Office of Energy Efficiency and Renewable Energy. *Battery Basics.*

- Standard lithium-ion EV batteries typically achieve 1,500–2,500 charge cycles.
- Lithium iron phosphate (LFP) batteries can achieve 6,000–10,000 cycles.

Electric Vehicle Adoption

U.S. Department of Energy, Alternative Fuels Data Center. *Electric Vehicle Registration Data.*

- EV registration reached 5% of new vehicle sales in the United States.

Critical Minerals

International Energy Agency. *The Role of Critical Minerals in Clean Energy Transitions*. Paris: IEA, 2021.

- Clean energy technologies require significantly more minerals than current recycling can supply.
- The transition will require approximately 40 times more minerals than current supply chains provide.

Additional Context

Several interviewees reference their own published work, industry reports, or research findings. These references are embedded within their narratives as part of their professional expertise. Examples include:

- Industry market share data (The Solar Installer)
- Program evaluation methodologies (The Program Designer, The Program Evaluator)
- Regulatory filings and testimony
- Company-specific metrics and case studies

These internal references reflect the speakers' professional knowledge and are presented as their accounts of their work, consistent with the oral history methodology.

Index

A
Agricultural runoff, 83
Alaska, 45, 46
Anonymization, 3
Anti-union industry, 16
Architecture, 50, 113, 136

B
Baseline methodology, 44, 80
Battery storage, 25, 124, 132, 143
Behavior change, 66
Behavior research, 49–51, 154
Benchmarking, 150
Blog publishing, 95, 96, 99, 101
Building decarbonization, 145
Building energy, 50, 65, 103, 129

C
Campus organizing, 59–61
Capacity building, 2
Carbon footprint, 49, 101
Career pathway, 2
Charging infrastructure, 117, 119
Clean Water Act, 83, 84
Climate activism, 73
Climate adaptation, 66
Climate careers, 1, 2
Climate grief, 73
Climate justice, 71
Climate skeptics, 151
Coalition building, 64
Community benefits, 27
Community energy, 100, 137
Community engagement, 26
Community solar, 20
Cost-effectiveness, 64, 77, 109, 112
Crane operations, 11
Customer education, 119

D
Data science, 53–55
Deferred maintenance, 131
Demand flexibility, 97
Demand response, 49, 76, 97, 105, 127
Digital twins, 44
Distributed energy, 135
Distributed generation, 19, 127
Diversity, equity, inclusion, 68
Dubai, 136

E
Egypt, 50
Electric scooters, 100, 101
Electric vehicles, 22, 117
Electrification, 2, 105, 127, 145
Embedded evaluation, 112
Emissions reduction, 35
Endangered species, 26
Energy access, 79–81
Energy consulting, 99

A. Kelly, *Good Work*, Women in Engineering and Science,
https://doi.org/10.1007/978-3-032-27547-9

Energy efficiency, 1, 35, 53, 76, 79, 80, 87, 88, 97, 104, 105, 107, 108, 113, 127–129, 131, 132, 135–137, 145, 146
Energy efficiency software, 53
Energy policy, 99
Energy research, 68
Energy sector, 88
Energy transition, 69, 96, 97
Entrepreneurship, 95
Environmental justice, 68, 76–77, 120
Environmental law, 72
Environmental microbiology, 45
Environmental permitting, 20
Environmental, Social, Governance (ESG), 93, 95, 96, 149, 150
Equity, 72, 76, 91, 92, 112, 117, 120, 123, 145
ESG consulting, 149–151
ESG funds, 96
EV charging, 63, 97
EV policy, 63–64

F
Facilities management, 154
First-generation students, 67
Forest conservation, 71
Freelance consulting, 112

G
Gamification, 49, 50
Generational equity, 151
Geological engineering, 39
Glacial research, 45
Graduate training, 36
Great Plains, 3, 11–13, 29
Green building, 136
Green jobs, 109
Grid interconnection, 25
Grid resilience, 123

H
Heat pumps, 146
Hydrogen energy, 93

I
Industrial decarbonization, 129
Industrial efficiency, 127–130
Iran, 111

K
Kentucky, 151

L
Labor, 8, 69, 128
Land use, 26
Leadership development, 107
Lifecycle analysis, 99
Lithium-ion batteries, 21

M
Machine learning, 44, 53–55
Manufacturing, 129
Marketing communications, 119
Measurable impact, 65
Measurement and verification, 53–55
Mentorship, 37, 110
Methane, 44, 46, 91–94
Methane capture, 91–94
Microgrids, 3, 123, 124
Microwave technology, 40
Mid-Atlantic, 99–102
Midwest, 19–20, 35–37, 39–41, 83–84, 127–130, 141–143
Military veterans, 153
Minimum energy performance standards, 79, 80
Mining byproducts, 40
Mountain West, 15–17, 87–94

N
National laboratory, 67
Natural gas, 26, 91, 93, 103
Nonprofit, 1, 71, 84, 109

O
Oral history methodology, 161

P
Pacific Northwest, 3, 25, 45–47, 53–55, 59–61, 63–64, 71–73, 97, 103–112, 117–118, 123–124, 131, 135–137, 145
Pay for performance, 53
Peer-review, 65, 66
Permitting, 20, 26, 64
PhD research, 49

Philanthropy, 79–81
Pipeline operations, 91
Power purchase agreements, 8, 26
Preventive maintenance, 8, 30
Process evaluation, 88
Program design, 76, 87, 88, 107, 108, 129
Program evaluation, 103
Public buildings, 131–133
Public speaking, 59

R
Range anxiety, 120
Range confidence, 120
Refrigerants, 79
Regenerative agriculture, 84
Regulatory barriers, 123
Regulatory compliance, 84
Regulatory proceedings, 63
Regulatory testimony, 161
Renewable natural gas, 91, 93
Residential solar, 15

S
Servant leadership, 135, 136
Skepticism, 37, 120
Social marketing, 65, 66
Software system, 142
Solar energy, 19, 96
Solar installation, 21
Southeast, 79–81, 149
Southwest, 75–77
Stakeholder engagement, 76, 88
Strategic energy management, 108, 128, 129, 135
Supplier diversity, 108
Sustainability data, 149, 150
Sustainable investing, 95, 96
Sustainable mining, 39

T
Team management, 136
Technical education, 54
Technical training, 23
Texas winter storm, 93
Transportation electrification, 63, 64, 117, 120
Transportation policy, 117
Tribal partnerships, 69
Turbine maintenance, 12
Turbine repair, 142

U
Union apprenticeship, 13, 153
Utilities commission, 63
Utility partnership, 75–77
Utility policy, 145–147
Utility regulation, 63, 88
Utility-scale renewables, 90

V
Virtual batteries, 96
Virtual fencing, 84

W
Watershed management, 83
Weathercasters, 65, 66
West Coast, 7–9, 21–23, 61, 93, 95–98, 119–121, 135
Western United States, 25
Wildfires, 22, 35, 72, 149
Wind farm management, 29–31
Wind technician, 142
Women in energy, 108
Women-owned business, 112
Workforce development, 67
Workforce training, 110
Working class, 7, 129

FSC
www.fsc.org
MIX
Papier | Fördert gute Waldnutzung
FSC® C083411

Kolibri 360 GmbH
Ferdinand-Jühlke-Straße 7
99095 Erfurt, Deutschland
produktsicherheit@kolibri360.de